Air Power in the New Counterinsurgency Era

The Strategic Importance of USAF Advisory and Assistance Missions

Alan J. Vick, Adam Grissom, William Rosenau, Beth Grill, Karl P. Mueller

Prepared for the United States Air Force

Approved for public release; distribution unlimited

The research described in this report was sponsored by the United States Air Force under Contract F49642-01-C-0003. Further information may be obtained from the Strategic Planning Division, Directorate of Plans, Hq USAF.

Library of Congress Cataloging-in-Publication Data

Air power in the new counterinsurgency era : the strategic importance of USAF advisory and assistance missions / Alan J. Vick ... [et al.].
p. cm.
"MG-509."
Includes bibliographical references.
ISBN-13: 978-0-8330-3963-7 (pbk. : alk. paper)
1. Counterinsurgency—United States. 2. Air power—United States. 3. United States. Air Force. 4. Military assistance, American 5. Military missions. 6. World politics—21st century. I. Vick, Alan.

U241.A57 2006
358.4'1425—dc22

2006019803

Cover image of CT team: Combat Aviation Advisors from the 6th SOS and regular army soldiers from Chad in front of a Chad Air Force C-130.

Published 2006 by the RAND Corporation
1776 Main Street, P.O. Box 2138, Santa Monica, CA 90407-2138
1200 South Hayes Street, Arlington, VA 22202-5050
4570 Fifth Avenue, Suite 600, Pittsburgh, PA 15213
RAND URL: http://www.rand.org/
To order RAND documents or to obtain additional information, contact
Distribution Services: Telephone: (310) 451-7002;
Fax: (310) 451-6915; Email: order@rand.org

Preface

With insurgency growing in importance as a national security problem, it is receiving new interest across the services, in the Department of Defense (DoD), and elsewhere in the U.S. government. Although ongoing operations in Afghanistan and Iraq give particular immediacy to the problem, the challenge of insurgency extends well beyond these specific conflicts. It is important, therefore, that the U.S. Air Force (USAF) consider how to meet the growing demand for air power in joint, combined, and interagency counterinsurgency operations and that other services' and DoD-wide reassessments of the subject take the potential roles of air power in counterinsurgency fully into account. In particular, airmen should take the lead in exploring how air power might work in combination with other military and civil instruments to help avert the development of an insurgency or perhaps to check a growing insurgency long enough to allow political and social initiatives (the heart of any successful counterinsurgency strategy) to take hold.

To address these and related policy challenges, RAND Project AIR FORCE conducted a fiscal year 2005 study entitled "The USAF's Role in Countering Insurgencies." This study addressed four major policy questions: (1) What threat do modern insurgencies pose to U.S. interests? (2) What strategy should the United States pursue to counter insurgent threats? (3) What role does military power play in defeating insurgencies? (4) What steps should USAF take to most effectively contribute to counterinsurgency? This work builds on more than 40 years of RAND Corporation work on insurgency, peace operations, and other types of lesser conflicts.[1]

This monograph has several purposes and audiences. First, it seeks to be a short primer on the problem of insurgency, counterinsurgency principles, and the role of air power in countering insurgencies. It is hoped that it will be a valuable introduction for airmen new to the topic. Second, it is hoped that the analysis on the potential demand for advisory assistance, as well as the data collection and analysis of recent 6th Special Operations Squadron (6 SOS) missions, will offer new insights to counterinsurgency practitioners in USAF. Finally, the monograph seeks to offer senior USAF leaders a way ahead to develop increased capability in this area without sacrificing the Air Force's edge in major combat operations.

The research reported here was sponsored by the Director of Operational Planning, Headquarters U.S. Air Force, and conducted within the Strategy and Doctrine Program of RAND Project AIR FORCE.

RAND Project AIR FORCE

RAND Project AIR FORCE (PAF), a division of the RAND Corporation, is the U.S. Air Force's federally funded research and development center for studies and analyses. PAF provides the Air Force with independent analyses of policy alternatives affecting the development, employment, combat readiness, and support of current and future aerospace forces. Research is conducted in four programs: Aerospace Force

---

[1] Between 1958 and 2005, RAND published over 50 reports with *counterinsurgency* in the title. In the same period, the abstracts for over 200 RAND reports included the term. For an overview of RAND work on this topic, see Austin Long, *On "Other War": Lessons from Five Decades of RAND Counterinsurgency Research*, Santa Monica, Calif.: RAND Corporation, MG-482-OSD, 2006. One of the earlier RAND works on counterinsurgency reports the results of a 1962 symposium at which scholars, planners, and practitioners came together to discuss the state of the art. See Stephen T. Hosmer and S. O. Crane, *Counterinsurgency: A Symposium, April 16–20, 1962*, Santa Monica, Calif.: RAND Corporation, R-412-ARPA, 1962. For a more recent work, see Bruce Hoffman, *Insurgency and Counterinsurgency in Iraq*, Santa Monica, Calif.: RAND Corporation, OP-127-IPC/CMEPP, 2004a.

Development; Manpower, Personnel, and Training; Resource Management; and Strategy and Doctrine.

Additional information about PAF is available on our Web site at http://www.rand.org/paf.

Contents

Figures

Tables

Summary

Often treated by Americans as an exceptional form of warfare, insurgency is anything but. Spanning the globe, centuries, and societies, insurgency is quite common. The United States itself was founded by insurgents—British colonists who rebelled against the abuses and neglect of British rule. At the end of the 19th century, the United States fought Filipino insurgents in its newly won territory. During the 20th century, U.S. forces fought insurgents in Nicaragua; Haiti; the Dominican Republic; the Philippines (again); Vietnam; and most recently, Afghanistan and Iraq. It has provided support to counterinsurgent forces in many more locations and support to insurgents in a few (most notably Nicaragua and Afghanistan).

This monograph seeks to help USAF prepare for future insurgency challenges by describing current trends, presenting an overview of key counterinsurgency principles, exploring counterinsurgency grand strategy options for the United States, proposing a new precautionary approach to counterinsurgency, and assessing current and potential USAF contributions.

Key Findings

- The primary insurgent threat to the United States today stems from regional rebels and global terrorists who share a common ideology.[1] These ties allow global terrorists to use a local insur-

[1] Throughout this report we use *insurgent* and *rebel* interchangeably.

gency as a training ground, to provide sanctuary, and to motivate a global audience (pp. 3–4).

- Today the only terrorist group with both the capability and desire to conduct attacks against U.S. interests at home and abroad is al Qaeda (pp. 24, 55).
- The U.S. counterinsurgency priority, therefore, should be the insurgencies motivated by radical Islam and global jihad. These are the ones most likely to find common cause with al Qaeda (p. 58).
- Previous experience with insurgencies has demonstrated that insurgencies are rarely defeated by outside powers. Rather, the best role for outsiders is an indirect one: training, advising, and equipping the local nation, which must win the war politically and militarily (pp. 4–5).
- A precautionary strategy that seeks to defeat the insurgency in its early stages is the most cost-effective approach, potentially avoiding huge costs in lives and dollars (pp. 82–93).
- Because insurgencies are fundamentally driven by social, political, and economic issues, nonmilitary aid will often be most important, especially in the early phases of a rebellion. Support to the host nation's police, security, and intelligence organizations is especially critical and should precede or occur in parallel with military assistance. When they are necessary, military actions must be carefully designed to support the overall political strategy. In past insurgencies, ill-considered actions by the government's military and security forces often increased insurgent determination and popular support for the rebels. The United States and leaders of partner nations must take care lest the military dynamic overtake the political (pp. 37–41, 45–47).
- Because air power has much to contribute to counterinsurgencies around the globe, advising, training, and equipping partner air forces will be a key component of U.S. counterinsurgency efforts (pp. 109–114).
- USAF needs a full-spectrum counterinsurgency capability. Although training, advising, and equipping efforts will be USAF's most common role in counterinsurgency, some situations may

require U.S. combat air power to team with indigenous or coalition ground forces or to participate in joint and interagency U.S. counterinsurgency operations (pp. 146–147).

Recommendations for USAF

USAF possesses a broad range of capabilities, in both its special and general-purpose forces, that can make significant contributions to fighting insurgents. Bringing these capabilities to bear on the counterinsurgency problem will require that counterinsurgency be treated as a problem as important as conventional warfighting, even though the manpower, dollars, and force structure devoted to it will likely never need to be as large as those devoted to major combat operations. To enhance its contribution to counterinsurgency, USAF should take the following steps:

- **Make counterinsurgency an institutional priority.** Without clear signals from senior USAF leaders, the institutional USAF will continue to treat counterinsurgency either as something that only the Air Force Special Operations Command (AFSOC) does or as a lesser included case that requires no special preparation. Major speeches, vision statements, personnel policy changes, and new programs will be necessary to overcome this perception (p. 133).
- **Create organizations and processes to oversee USAF counterinsurgency efforts.** The USAF will need new organizations to develop and oversee counterinsurgency policy and concepts, to integrate efforts across the USAF, to coordinate with DoD and other agencies, and to execute counterinsurgency advisory and assistance missions (pp. 133, 135).
- **Develop and nurture counterinsurgency expertise throughout USAF.** Counterinsurgency expertise does exist in USAF, but, outside of AFSOC, it is scattered and limited. Substantial counterinsurgency education should be a mandatory part of the cur-

riculum in the Air Force Reserve Officer Training Corps, at the Air Force Academy, and in all phases of Air Force Professional Military Education from Squadron Officer School to the Air War College. Opportunities for more in-depth training and education will need to be developed, as will appropriate career paths for counterinsurgency specialists (pp. 135–136).

- **Create a wing-level organization for aviation advising.** This is likely the single most important initiative USAF can take to enhance its own counterinsurgency capabilities. By creating a wing-level organization, USAF will be able to grow its advisory capacity to meet the demand; expand aviation assistance to include institutional and higher-level advising; develop new counterinsurgency concepts and technologies for partner air forces; supervise an embedded advisor program; and offer sufficiently diverse opportunities to attract and retain the very best officers, noncommissioned officers, and civilian personnel (pp. 136–143).
- **Enhance USAF combat capabilities for counterinsurgency.** Although only as a last resort, USAF does need the ability to conduct air operations in support of partner-nation forces and/or U.S. joint forces fighting insurgencies. USAF already has considerable relevant capabilities, and its modernization programs will enhance them further. Beyond that, specific technologies (e.g., foliage-penetrating sensors) and, most important, a deeper understanding of the insurgent phenomenon will increase the effectiveness of air power in future counterinsurgency operations (pp. 146–147).

Acknowledgments

Maj Gen Roy M. Worden, Director for Operational Plans and Joint Matters, Headquarters, USAF, was the study sponsor. We greatly appreciate General Worden's enthusiastic support, his assistance in thinking through the implications of a precautionary strategy, and his efforts to ensure that the study recommendations reach key decisionmakers.

Lt Col Thomas McCarthy, HQ USAF/XOXS, was the study action officer. We benefited enormously from his expert insights and recommendations, as well as his careful attention to more-mundane administrative matters.

We thank Andrew Hoehn, Director of PAF's Strategy and Doctrine Program, for arranging two meetings with senior military leaders who are intensively involved in military assistance activities in key regions. Gen Charles Wald, USAF, Deputy Commander European Command, and Lt Gen Wallace Gregson, USMC, Commander Marine Forces Pacific, took time out of their busy schedules to discuss the challenges of foreign internal defense in their respective theaters.

We thank Lt Gen Jeffrey Kohler, Director, Defense Security Cooperation Agency; Lt Gen Michael Wooley, Commander, AFSOC; and Maj Gen John Folkerts, Vice Commander, AFSOC, for their helpful comments on the project briefing.

The study team received exceptional support and assistance from the leadership and personnel of the 6th Special Operations Squadron (6 SOS), Hurlburt Field, Florida. We thank Jerome Klingaman, Director of Strategy and Plans, 6 SOS, for his encouragement, gracious hospitality, and generosity in sharing insights he gained as a combat aviation advisor over the last four decades. We also thank Lt Col Juan

Alvarez, Commander, 6 SOS, for supporting our multiple visits and for his helpful comments on our briefing. SMSgt Hale Laughlin also shared his expertise and insights with us on multiple visits. Diane Beck, Program Analysis and Documentation, provided essential help gaining access to 6 SOS after-action reports and other documents.

AFSOC hosted multiple visits by our research team to Hurlburt Field. We thank Colonel Norman Brozenick, Commander, 16th Special Operations Wing, for his helpful comments on our briefing and for sharing his experiences as a former commander of the 6 SOS and as a combat aviation advisor. Lt Col James Walker, Todd Kratzke, and Lt Col Craig Werenskjold helped organize study team meetings with personnel from the 16th SOW operational squadrons. Lt Col Timothy Finnegan, Lt Col Adam Mlot, SMSgt Ken Graff, and Terrence Sykes all met with us to discuss foreign internal defense issues. Herb Mason, AFSOC Historian, kindly helped the project team identify and gain access to relevant AFSOC historical materials.

Finally, we thank the aircrews from the 4th, 8th, 9th, 15th, 16th, and 20th SOSs, who shared their recent combat experiences in Afghanistan and Iraq.

Caesar Sereseres, Associate Dean of Undergraduate Studies, School of Social Sciences, University of California, Irvine, shared his considerable knowledge on U.S. assistance to El Salvador in the 1980s and offered his perspective on modern insurgencies. Andrea Lopez, Professor of Political Science, Susquehanna University, shared her insights on counterinsurgency operations in Afghanistan.

Greg Jannarone, Chief, Behavioral Influences Analysis Division, National Air and Space Intelligence Center, Wright Patterson AFB, hosted a study team visit. We thank Mr. Jannarone and division personnel for their constructive comments on our briefing and for a fascinating discussion of the human factors side of the insurgency problem.

Lt Col David Kilcullen, Royal Australian Army, met with project members to share his experiences in counterinsurgency operations and kindly provided access to his various manuscripts and publications on insurgency.

Maj Yvette Quitno, Headquarters USAF, provided a helpful alternative perspective on how to organize advising and training activities.

Maj Gen Jonathan Gration (USAF), Col Thomas Griffith (USAF), Colonel Dennis Jones (USAF), Lt Col Adam Mlot (USAF), Lt Col Michael Gendron (USAF), Col Robyn Read (USAF, ret.), and Diane Beck (USAF) all provided helpful comments and suggestions on the draft report.

RAND colleagues Natalie Crawford, Edward Harshberger, Andrew Hoehn, Stephen Hosmer, Jefferson Marquis, Forrest Morgan, Melinda Moore, Jennifer Moroney, Bruce Nardulli, David Ochmanek, Olga Oliker, Bruce Pirnie, James Quinlivan, Angel Rabasa, David Shlapak, Steven Simon, Michael Spirtas, and David Thaler provided valuable comments and suggestions on earlier versions of this work. Albert Robbert helped us understand the USAF pilot bonus program and assisted us in developing the manpower metric for aviation advisors. Rollie Lal discussed her work on international crime and terrorism links with our study team. Nora Bensahel served on the project team and wrote a separate paper on the challenge of insurgencies following regime change. Rob Owen, Professor of Aeronautical Science, Embry-Riddle University and an adjunct member of the RAND staff, contributed to project deliberations on insurgency and wrote a separate paper on the role of airlift during insurgencies. RAND Summer Associate Christopher Darnton kindly shared his insurgency database and trends analysis with our project team.

Bruce Hoffman of RAND and Max Manwaring, Professor of Military Strategy, U.S. Army War College, were the formal reviewers of this report. We thank them for their constructive and thoughtful critiques.

Natalie Ziegler prepared the manuscript and provided excellent administrative support to the study team.

Finally, we thank Phyllis Gilmore, our exceptional editor, for her many contributions to this report.

Abbreviations

6 SOS	6th Special Operations Squadron
AFSOC	Air Force Special Operations Command
BCE	before the common era
BNCOC	Basic Noncommissioned Officer Course
CENTCOM	U.S. Central Command
CONOP	concept of operation
CRS	Congressional Research Service
DoD	Department of Defense
ESAF	El Salvador Armed Forces
EUCOM	U.S. European Command
FARC	Fuerzas Armadas Revolucionarias de Colombia [Revolutionary Armed Forces of Colombia]
FAS	Salvadoran Air Force
FMLN	Farabundo Marti Liberation Front
FMF	foreign military financing
FY	fiscal year
GTEP	Georgia Train and Equip Program
IMATT	International Military and Advisory Training Team
IMET	International Military Education and Training
JCET	Joint Combined Exchange Training

LTTE	Liberation Tigers of Tamil Eelam (also known as the Tamil Tigers)
MTT	mobile training team
NATO	North Atlantic Treaty Organization
NCO	noncommissioned officer
OAD	operational aviation detachment
OIF	Operation Iraqi Freedom
PACAF	Pacific Air Force
PACOM	U.S. Pacific Command
PAF	RAND Project AIR FORCE
PCS	Communist Party of El Salvador
RSLAF	Republic of Sierra Leone Armed Forces
SOCOM	Special Operations Command
SOF	special operations forces
SOUTHCOM	U.S. Southern Command
TDY	temporary duty
UAV	unmanned aerial vehicle
USAF	U.S. Air Force

CHAPTER ONE
Introduction

Background

Insurgency is not a new form of warfare, dating back to at least 165 BCE when insurgent Jews under Judas Maccabeus defeated Greek occupiers and liberated Jerusalem.[1]

Neither is insurgency new to the United States. The U.S. military has either fought insurgents or supported friendly governments in many counterinsurgency operations since the early 20th century. The Philippines, Haiti, Nicaragua, Greece, Vietnam, Thailand, Laos, Cambodia, Bolivia, El Salvador, Colombia, Afghanistan, and Iraq are only the most prominent examples.[2] During the 40 years of the Cold War, the United States actively sought—through economic aid, security assistance, and combat operations—to counter communist insurgen-

---

[1] The Central Intelligence Agency defines *insurgency* as

> protracted political-military activity directed toward completely or partially controlling the resources of a country through the use of irregular military forces and illegal political organizations . . . The common denominator of most insurgent groups is their desire to control a particular area. This objective differentiates insurgent groups from purely terrorist organizations, whose objectives do not include the creation of an alternative government capable of controlling a given area or country.

See Central Intelligence Agency, *Guide to the Analysis of Insurgency*, Washington, D.C.: U.S. Government Printing Office, no date.

[2] See Anthony James Joes, *America and Guerrilla Warfare*, Lexington, Ky.: University Press of Kentucky, 2000.

cies around the world.[3] These insurgencies were seen as part of a global communist strategy to spread instability, install Marxist governments, undermine democracy, and isolate the United States and other Western powers. U.S. government attention to counterinsurgency peaked during the Vietnam War. When that ended, the defense community rapidly shifted its attention back to the twin threats posed by Soviet nuclear and conventional forces. Although the Reagan administration committed significant resources to opposing insurgencies (as well as supporting several) during the 1980s, the military services remained largely indifferent toward the problem. Writing in 1988, Dennis Drew observed that

> the American military has all but turned its back on the study and preparation for low-intensity conflicts and has concentrated its efforts on worst case scenarios involving nuclear deterrence and a major war against the Warsaw Pact in Europe or Southwest Asia.[4]

The end of the Cold War only exacerbated this trend, largely ending official Washington interest in the civil war in El Salvador, for example. To the extent that insurgency mattered to U.S. security policy, it was limited to those in a few key countries, such as Colombia and the Philippines. A small cadre of insurgency specialists survived in the special operations and intelligence worlds, academia, and think tanks, but the broader defense community quickly lost sight of counterinsurgency as a military challenge.

Since September 11, 2001, however, the problem of insurgency has once again become a priority for the U.S. government, largely

---

[3] The 1960s marked the height of U.S. interest in counterinsurgency. Although the bulk of U.S. efforts were consumed by the conflict in Southeast Asia, there was considerable activity elsewhere. For example, between 1962 and 1968, the 8th U.S. Army Special Forces Group, based in Panama, conducted over 400 internal security–related missions in Latin America alone. See Ian F. W. Beckett, *Modern Insurgencies and Counter-Insurgencies: Guerrillas and Their Opponents Since 1750*, London, UK: Routledge, 2001, p. 173.

[4] Dennis Drew, *Insurgency and Counterinsurgency: American Military Dilemmas and Doctrinal Proposals*, Maxwell AFB, Ala.: Air University Press, March 1988, p. 1. See also Andrew J. Bacevich et al., *American Policy in Small Wars: The Case of El Salvador*, Washington, D.C.: Pergamon-Brassey's, 1988, especially pp. 14–15.

because of the connection between Islamic insurgents and global jihadist groups, such as al Qaeda. With ties among insurgent and terrorist groups expanding, the line between global counterterrorist actions and counterinsurgency is becoming blurred. The United States is currently conducting counterinsurgency operations or providing support to governments facing insurgencies in Afghanistan, the Philippines, Colombia, Georgia, Iraq, and elsewhere. Among the instances of major U.S. involvement, there are significant ties between local insurgents and global jihadists in all but Colombia.

In Iraq, the United States is learning once again that counterinsurgency operations are complex, dangerous, difficult, and time consuming. Although the Iraq experience is unique in some respects, it is a powerful reminder of some common elements all insurgencies share. In particular, successful counterinsurgency requires tight integration of political, military, intelligence, police, and economic activities and organizations—a feat that is inherently difficult. It also requires that their actions be well integrated with those of the local government and of any other states, alliances, or other multinational organizations participating in the intervention.[5] Although the U.S. military can achieve rapid and operationally decisive outcomes in conventional conflict, it has been less successful against insurgents, and, in any event, the military instrument can play only a comparatively small, if nevertheless essential, role in defeating an insurgency. That said, there may be situations in which U.S. military forces do need to intervene to help stabilize a situation so that the local government can address the roots of the insurgency and build up its own security capabilities.

Whether the United States achieves its goals in Iraq or not, the experience there should not mask a fundamental truth: The nexus between local insurgencies and terrorist groups with global ambitions means that the United States can ignore insurgencies only at its own peril. Insurgent groups that control territory, are involved in smuggling, or possess military or other skills can provide significant support to global terrorists. The existence of Islamic insurgencies is also

---

[5] Counterinsurgency is likely to also require the United States to work effectively with the United Nations, other international organizations, and nongovernmental organizations.

enormously helpful, if not essential, for global jihadism because they motivate and inspire a global audience, help recruiting and fund raising, and can provide a crucible for testing and training new recruits. At the same time, connections to terrorist groups tend to increase both the level of hostility toward the United States and the capabilities of local insurgent groups to challenge state authority and attack U.S. interests. Although not every insurgency will be a potential threat to U.S. interests, many will require carefully calibrated U.S. action. In other cases, successfully assisting other states in their counterinsurgency operations may help avert the emergence of threats to U.S. national security over the longer term.

The Dilemma of Intervention

Because the United States is an outside power intervening in what locals may view as an internal matter, any U.S. involvement always carries the seeds of its own defeat. The very presence of U.S. forces, particularly those involved in combat operations, may stir opposition, be perceived as part of a broader design to support U.S. hegemony, or be viewed as supporting an illegitimate local government. This is especially so in regions where the United States (because of its policies, past actions, or culture) is viewed with suspicion or hostility. Even tactical victories may be operational defeats when the deaths of insurgents and, especially, noncombatants in combat operations motivate others to join the struggle. To the extent that the local populace identifies with a larger movement (e.g., global jihadism), U.S. policies elsewhere may undermine local support for a friendly government. In short, external involvement in insurgencies is fraught with complex and paradoxical dynamics. If the United States is going to be successful in defeating threatening insurgencies, it will need to develop a broad strategy that is sensitive to these risks and mixes military, law enforcement, intelligence, and other instruments of power to undermine and ultimately end support for the insurgents.

The fundamental goal in any counterinsurgency operation must be to gain the allegiance of the population to the government.[6] Everything that the local government, the United States, and other participants do must be assessed in light of the contribution to this goal. In general, outsiders contribute to this fundamental goal only indirectly. Police, military, intelligence, economic, and other assistance may be essential to strengthen a government fighting insurgents, but, by themselves, they do not directly contribute to this goal. For example, U.S. civic-action programs (e.g., digging wells, building schools) are often greatly appreciated by the local populace and may enhance U.S. standing but are not likely to enhance allegiance to the central government. Indeed the U.S. power, enthusiasm, and competence displayed in such activities is often in such stark contrast to the performance of their own government that it may further undermine allegiance. At best, the United States may be able to use civic action to build friendships and gain allies who will work with the United States to fight the insurgents, but that is a temporary measure at best. Ideally, the focus of all U.S. activities would be to give the partner government the resources and training so that it could take the political, military, economic, and other initiatives that would convince the people that the government is worthy of their allegiance.

Given these constraints on outside intervention, this monograph emphasizes the role of the U.S. military, and USAF in particular, in training, advising, and equipping partner nations so that they can successfully deal with insurgencies.[7] The precautionary strategy we discuss here is consistent with recent DoD moves to take an indirect approach to battling insurgents and terrorists, emphasizing building partner capabilities rather than direct combat operations by U.S. forces.[8]

---

[6] Thanks to RAND colleague Bruce Pirnie for sharing his insights on how U.S. activities in Afghanistan and Iraq might be conducted to better support this objective.

[7] Although the emphasis here is on military assistance, we recognize that support to the host nation's police, security, and intelligence organizations is especially critical and should precede or occur in parallel with military assistance.

[8] See U.S. Department of Defense, *Quadrennial Defense Review Report*, Washington, D.C., February 6, 2006, especially pp. 2 and 87–91.

Purpose and Organization of This Monograph

The objective is to help USAF explore its potential role in future counterinsurgency operations. In particular, we address four major policy questions: (1) What threat do modern insurgencies pose to U.S. interests? (2) What strategy should the United States pursue to counter insurgent threats? (3) What role does military power play in defeating insurgencies? (4) What steps should USAF take to most effectively contribute to counterinsurgency?

Chapter Two explores how the insurgency phenomenon has evolved and the nature of the current challenge to the United States. Chapter Three presents lessons learned from counterinsurgency over the last 60 years or so. Chapter Four approaches the problem from the level of grand strategy, assessing the types of military capabilities and strategies necessary to deal with the counterinsurgency challenge. Chapter Five discusses the advantages of a precautionary strategy that seeks to head off insurgencies while they are still quite young. Chapter Six assesses USAF's current contributions in counterinsurgency and explores options to enhance this role in the future. Chapter Seven presents our conclusions and recommendations for USAF. Appendix A contains additional information on current insurgencies. Appendix B explains the derivation of the manpower metric presented in Chapter Six.

CHAPTER TWO

The Evolving Insurgency Challenge

Introduction

After a decade or more of languishing in obscurity, the phenomenon of insurgency reemerged as a subject of official, analytical, and academic interest during the first years of the 21st century. The ongoing conflicts in Iraq and Afghanistan illustrate in dramatic terms that insurgency can pose a considerable challenge for even the most formidable military power. Increasingly, the Bush administration defined these insurgencies as fronts within a burgeoning global Islamist insurgency that includes, but is not limited to, Osama bin Laden's network.[1] Indeed, as bin Laden declared in an October 2004 speech, the techniques of insurgency, honed in the 1980s during the struggle against the Soviet occupation of Afghanistan, are now being employed elsewhere as part of a global drive "to make America bleed profusely to the point of bankruptcy."[2]

Insurgency as form of political-military struggle is not new. Guerrillas and partisans have existed throughout recorded history, although irregular conflict did take on a new form in the 20th century, when "social, economic, psychological, and, especially, political elements [were] grafted onto [guerrilla] tactics in order to radically alter the

---

1 See for example U.S. Agency for International Development, "USAID's Role in the War on Terrorism," Issue Brief 1, 2001.

2 Middle East Media Research Institute, "The Full Version of Osama bin Laden's Speech," Special Dispatch Series, No. 811, November 5, 2004.

structure of the state by force."[3] The strategy, which employs limited means in unconventional, ruthless, and audacious ways to pursue what J. Bowyer Bell termed "maximum ends,"[4] remains a popular one, as evidenced by the number and duration of current insurgencies.[5]

As always, conceptual clarity is a critical first step in fashioning an effective response to national security challenges. Toward that end, this chapter will focus on several aspects of the contemporary insurgency phenomenon that civilian and military planners might usefully consider as they devise approaches for protecting U.S. interests in the new international security environment. After a discussion of definitions, the chapter will consider efforts to categorize insurgent movements. It will then explore broad trends in insurgency and evaluate factors that are helping to create and sustain these conflicts. The chapter will conclude with a discussion of insurgent threats to U.S. interests, and will offer some caveats for policymakers to consider as they grapple with the challenge of waging counterinsurgency on a global scale.

Defining Insurgency

"Insurgency," like many political terms, is a contested concept.[6] It is therefore not surprising that there is no generally agreed on definition within the armed forces, the U.S. government, or the broader policy community. What is more, many of the definitions in wide circulation are problematical. The Department of Defense (DoD), for example, defines *insurgency* as "an organized movement aimed at the overthrow of a constituted government through use of subversion and armed

[3] Ian F. Beckett, *Insurgency in Iraq: An Historical Perspective,* Carlisle, Pa.: U.S. Army War College Strategic Studies Institute, January 2005, p. 2. For a survey of pre–20th century partisan warfare, see Lewis H. Gann, *Guerrillas in History*, Stanford, Calif.: Hoover Institution Press, 1971, Chapters 1 and 2.

[4] J. Bowyer Bell, *The Dynamics of the Armed Struggle*, London, UK: Frank Cass, 1988, p. 220.

[5] See Appendix A for a list of states afflicted by insurgencies.

[6] For more on this notion, see Michael Walzer, "Five Questions About Terrorism," *Dissent*, Winter 2002, p. 5.

conflict."[7] This definition has the advantage of parsimony, but fails to capture essential aspects of the phenomenon, such as the fundamentally political nature of the insurgency movement, and the centrality of the local population in any insurgency campaign. What is more, the current conflict in Iraq might not qualify as an insurgency under this definition, since the opposition forces there scarcely constitute an "organized movement" and the "constituted government," such as it was, initially took the form of the Coalition Provisional Authority. A widely cited academic definition describes *insurgency* as a "technology of military conflict characterized by small, lightly armed bands practicing guerrilla warfare from rural base areas. As a form of warfare insurgency can be harnessed to diverse political agendas, motivations, and grievances."[8] While highlighting some important features, such as the fact that a variety of different motivations can animate insurgents, this definition also falls short, since it seems to suggest, among other things, that insurgency is an exclusively rural phenomenon. For its part, the British Army defines insurgency as the "actions of a minority group within a state who are intent on forcing political change by means of a mixture of subversion, propaganda and military pressure, aiming to persuade or intimidate the broad mass of people to accept such a change."[9] This definition correctly stresses the central role of the local population, but it too falls short by failing to capture explicitly the fundamental notion that insurgency and counterinsurgency are in essence competitions for legitimacy.

Rather than belabor the point, it might be useful to consider another definition for what is obviously not a straightforward concept. Although hardly parsimonious, the definition Richard H. Shultz, Douglas Farah, and Itamara V. Lochard offer reflects the full range of the key political and operational components of insurgency:

---

[7] Joint Staff, *Department of Defense Dictionary of Military and Associated Terms*, April 12, 2001 (as amended through May 9, 2005).

[8] James D. Fearon and David D. Laitin, "Ethnicity, Insurgency, and Civil War," *American Political Science Review*, Vol. 97, No. 1, February 2003, pp. 75–90; see p. 75.

[9] [British] Army Field Manual, Vol. 1, *Combined Arms Operations*, Part 10, *Counter Insurgency Operations (Strategic and Operational Guidelines)*, July 2001, p. A-1-1.

> a protracted political and military set of activities directed toward partially or completely gaining control over the territory of a country through the use of irregular military forces and illegal political organizations. The insurgents engage in actions ranging from guerrilla operations, terrorism, and sabotage to political mobilization, political action, intelligence/counterintelligence activities, and propaganda/psychological warfare. All of these instruments are designed to weaken and/or destroy the power and legitimacy of a ruling government, while at the same time increasing the power and legitimacy of the armed insurgent group."[10]

This definition has a number of important strengths. It highlights insurgency as a struggle for power and legitimacy, stresses the range of violent and nonviolent instruments that insurgents typically employ, and presents the idea that insurgency is in part a psychological struggle aimed at eroding the incumbent power's will to continue the struggle. Most important, this definition helps us draw distinctions between insurgency and terrorism, a perennial conundrum for civilian and military officials, analysts, and journalists. Terrorists and insurgents share important features (e.g., their status as nonstate actors, their use of violence, and the clandestine nature of many of their activities), but these are outweighed by their differences, some of which are described in Table 2.1. Similarly, while sound counterterrorism strategies necessarily share many features with counterinsurgency—such as recognizing the importance of building local, "host nation" capacity—there are significant differences.[11] Historically, terrorists have not posed a threat to the survival of the state, as discussed below. Counterterrorism campaigns, therefore, seldom require the all-out mobilization of the entire apparatus of the state, as would be necessary to thwart a full-blown insurgent movement.

---

[10] Richard H. Shultz, Douglas Farah, and Itamara V. Lochard, *Armed Groups: A Tier-One Security Priority*, Colorado Springs, Colo.: USAF Institute for National Security Studies, Occasional Paper 57, September 2004, pp. 17–18. See also Central Intelligence Agency (no date).

[11] For more on this point, see David Ochmanek, *Military Operations Against Terrorist Groups Abroad: Implications for the United States Air Force*, Santa Monica, Calif.: RAND Corporation, MR-1738-AF, 2003, particularly Chapter Two.

Table 2.1
Key Differences Between Terrorists and Insurgents

	Terrorists	Insurgents
Targets	Primarily noncombatants	Primarily official (e.g., military, police, government personnel) and economic targets
Operations	Attacks carried out by members of small cells	Paramilitary and military, in larger formations
Territory	Rarely hold territory, and then only for short periods	Hold larger amounts of territory; some create "liberated zones"
Other	No uniforms; rarely if ever abide by the Law of Armed Conflict	Sometimes insurgents wear uniforms; sometimes respect Law of Armed Conflict

Terrorism, like insurgency, is a contested concept that is both a normative judgment and a descriptive term.[12] No "terrorists" want to be labeled as such. Rather, they would like to be categorized as they see themselves—as "guerrillas," "soldiers," "militants," or indeed, as "insurgents." But while all insurgents from time to time engage in terrorism—that is, the "deliberate creation and exploitation of fear through violence or the threat of violence in the pursuit of political change"[13]—no terrorists, their political rhetoric and propaganda notwithstanding, have a credible chance of achieving every insurgency's paramount goal: "partially or completely gaining control over the territory of a country." While many terrorists have reached short-term objectives, such as publicity for their cause, no government has ever been brought down as a result of terrorist actions.[14] Among other things, no purely terror-

---

[12] Charles Townsend, *Terrorism: A Very Short Introduction*, Oxford: Oxford University Press, 2002, pp. 1–8, and Bruce Hoffman, *Inside Terrorism*, New York: Columbia University Press, 1998, pp. 13–44.

[13] Hoffman (1998, p. 43). The U.S. Department of State defines terrorism as "premeditated, politically motivated violence perpetrated against noncombatant targets by subnational groups or clandestine agents, usually intended to influence an audience" (U.S. Department of State, Office of the Coordinator for Counterterrorism, *Patterns of Global Terrorism*, Washington, D.C., April 2003, p. xiii).

[14] Conor Gearty, *Terror*, London, UK: Faber and Faber, 1991, p. 2. Some would argue that the March 11, 2004, terrorist bombings of four commuter trains in Madrid, Spain, are an

ist movement has ever succeeded in mobilizing a sizeable population. On the contrary, terrorist violence typically alienates large segments of society, thereby creating a backlash that ultimately favors the state. In short, terrorists have never accomplished what the Front de Libération Nationale insurgents did in Algeria, the Khmers Rouges did in Cambodia, or the Tigray People's Liberation Front did in Ethiopia: driving out or destroying an incumbent power, seizing the apparatus of state control, and exercising authority over substantial physical territory.

Categorizing Insurgencies

During the Cold War, when U.S. policymakers saw insurgency as an instrument of Soviet (and, to a lesser extent, Chinese) foreign policy, the phenomenon was seen as an alarming manifestation of "warfare by other means." Conventional conflict and, with it, the potential for all-out nuclear war, drove the East-West struggle down to less dangerous but still perilous expressions, such as the so-called "brushfire" wars in Asia, Latin America, and Africa.[15] In the early 1960s, during the first "counterinsurgency era," and during the 1980s, when Soviet-backed revolutionary movements in Central America generated new interest

---

example of terrorist action bringing down a government. Prime Minister Jose Maria Anzar's Popular Party was expected to win reelection but was upset by the Socialists in the March 14 election. According to this perspective, Spanish voters turned against Anzar to achieve the withdrawal of Spanish forces from Iraq (and presumably to make Spain less of a target for terrorist attacks motivated by Spanish involvement in that conflict). Another view suggests that the election outcome was driven instead by voter anger over the government's clumsy handling of the investigation, particularly its rush to blame the ETA (Euskadi ta Askatasuna, the leading Basque separatist group) and the perception that it was withholding information from the public because the Popular Party feared it would hurt them on election day. The latter view does implicitly suggest a link (at least in the minds of the Popular Party leaders) between Spain's involvement in Iraq and the possibility that the terrorist attacks were retaliation and/or sought to undermine Spanish public support for the Popular Party's policies regarding Iraq. See Margaret Warner, "Aftermath in Spain," transcript of discussion with Richart Burt, Charles Kupchan, Daniel Benjamin, and Nicolas Checa, Public Broadcasting Service, *Newshour with Jim Lehrer*, transcript, March 15, 2004.

[15] Fred Halliday, *Revolution and World Politics: The Rise and Fall of the Sixth Great Power*, Durham, N.C.: Duke University Press, 1999, p. 247.

in defeating insurgencies, the U.S. armed forces, intelligence community, foreign assistance agencies, the Department of State, and other institutions devoted considerable resources to trying to understand and prepare for what was perceived to be a new and uniquely demanding set of challenges.

This Cold War experience left a legacy. Today, as the United States responds to serious insurgencies in Iraq and Afghanistan, as well as the emergence of what some policymakers and analysts have termed a "global" Islamist insurgency,[16] old patterns of thought retain their hold on our analytical imaginations. To be sure, many of the basic requirements for successful insurgency and counterinsurgency are essentially unchanged.[17] However, our frameworks for assessing and analyzing insurgent movements remain heavily colored by Western experiences during the Cold War and, in particular, the Vietnam War. Specifically, the Maoist strategy of protracted "popular war," and its Vietnamese variants, is the conceptual lens through which many officials, military officers, and journalists continue to view insurgent movements. To be sure, Maoist people's war is not obsolete. In South Asia, for example, the Communist Party of Nepal (Maoist) and India's "Naxalites" are employing the strategy, and with great success to date in the case of the former.[18] But the Maoist approach, with its emphasis on creating an alternative state, mobilizing a mass base, and employing a three-step political-military strategy culminating in a conventional open battle ("defensive," "equilibrium," and "offensive" phases, in Mao's formulation) is not in evidence today in Iraq or Afghanistan.[19] Yet the U.S. Army's current doctrine, in its discussion of insurgency,

---

[16] See, for example, John Mackinlay, *Globalisation and Insurgency*, London, UK: International Institute for Strategic Studies, Adelphi Paper 352, 2002, p. 79.

[17] Beckett (2005, p. 15)

[18] Rahul Bedi, "Maoist Activity Increases in India," *Jane's Intelligence Review* (online edition), April 2005, and Thomas A. Marks, *Insurgency in Nepal*, Carlisle, Pa.: U.S. Army Strategic Studies Institute, December 2003.

[19] Steven Metz and and Raymond Millen, *Insurgency and Counterinsurgency in the 21st Century: Reconceptualizing Threat and Response*, Carlisle Barracks, Pa.: U.S. Army Strategic Studies Institute, November 2004, p. 18.

continues to stress the insurgent goal of building a "counterstate" that will emerge from the shadows and assume power, as in China during the late 1940s.[20]

The "universalization" of the Maoist approach to insurgency may, however, cloud our thinking about what is required to neutralize, contain, or defeat insurgencies. As Donald Snow observed in 1996, "[i]t is not enough . . . to apply counterinsurgency doctrine developed to blunt the mobile-guerrilla strategy to the ongoing narco-insurgencies in Peru and Colombia."[21] Today, one would add Afghanistan and particularly Iraq to this list. In Iraq, the insurgency is being waged by "loose networks of state and nonstate actors, more like a social movement than the typical vertically organized guerrilla insurgency of earlier wars."[22] Iraq is in the midst of "an unusually invertebrate insurgency, without a central organization or ideology, a coherent set of objectives or a common positive purpose."[23] Indeed, the insurgency, made up of as many as 40 subgroups, according to some estimates, has multiple goals, including driving out the occupation forces, preventing the establishment of a liberal democracy, establishing Iraq as a base of jihadist operations, reinstituting Ba'athist rule, and personal enrichment. An in-depth exploration of counterinsurgency in Iraq is beyond the scope of this chapter.[24] However, it is reasonable to conclude that thwarting the insurgency will require an approach that moves beyond a counter–people's war strategy.

---

[20] U.S. Army, *Counterinsurgency Operations*, Washington, D.C.: Headquarters, Department of the Army, Field Manual—Interim 3-07.22, October 2004, p. 1-1.

[21] Donald M. Snow, *Uncivil Wars: International Security and the New International Conflicts*, Boulder, Colo.: Lynne Rienner Publishers, Inc., 1996, pp. 49–50.

[22] Mary Kaldor, "Iraq: The Wrong War," *Open Democracy*, September 6, 2005.

[23] Eliot A. Cohen, "A Hawk Questions Himself as His Son Goes to War," *Washington Post*, July 10, 2005, p. B1. For a different view, see Christopher Hitchens, "History and Mystery: Why Does the *New York Times* Insist on Calling Jihadists 'Insurgents'?" *Slate*, May 16, 2005.

[24] For an early and widely read assessment by a noted insurgency scholar, see Bruce Hoffman, *Insurgency and Counterinsurgency in Iraq*, Santa Monica, Calif.: RAND Corporation, OP-127-IPC/CMEPP, 2004a.

Creating typologies of insurgencies has been a perennial task for analysts since the early 1960s. Typically, these typologies focus on a single factor: motivation (Marxist, separatist, nationalist, etc.). Critics of this approach note that it ignores a broader set of operational and organizational characteristics, such as leadership, organization, training, and recruitment practices, which are likely to prove useful in fashioning a counterstrategy.[25] What is more, in overemphasizing motivation, we may be blinding ourselves to the fact that different groups might warrant different responses, despite the fact that they share a common typological label. Not all "nationalists" or "separatists" are alike, a reality that such labels might cause us to overlook. [26]Finally, we must always be alert to the fact that motivation is never uniform across an insurgent group.[27] While the leadership typically espouses an ideological agenda, the movement's rank and file are more often motivated by concrete, local grievances. This held true even during the Cold War, supposedly the high-water mark of ideologically driven insurgencies.[28] Put another way, "at the mass level, local considerations tended to trump ideological ones."[29]

Iraq is the latest in a series of painful reminders that insurgency takes a variety of forms, ranging from highly centralized, lethal, and professional, such as Lebanese Hizbollah, to the badly organized, ill disciplined, and largely incoherent, such as the Revolutionary United Front in Sierra Leone, a movement led by charismatic criminals whose political program took a back seat to its quest for spoils. What might be termed "legacy" insurgencies persist in Colombia and Peru, where drug trafficking has added a layer of complexity, criminality, and ideological confusion to what had been more-traditional Maoist-style strug-

---

[25] Mackinlay (2002, pp. 41–44).

[26] Mackinlay (2002, pp. 41–42).

[27] James C. Scott, "Revolution in the Revolution: Peasants and Commissars," *Theory and Society*, Vol. 7, No. 1/2, January–March 1979, pp. 97–134.

[28] Thomas A. Marks, "Ideology of Insurgency: New Ethnic Focus or Old Cold War Distortions?" *Small Wars and Insurgencies*, Vol. 15, No. 1, Spring 2004, p. 111.

[29] Stathis N. Kalyvas, "'New' and 'Old' Civil Wars: A Valid Distinction?" *World Politics*, Vol. 54, No. 1, October 2001, p. 107.

gles.[30] In Turkey, the Kurdistan Freedom and Democracy Congress, whose leadership espouses a Marxist-Leninist (but not Maoist) political program, is struggling to carve out an independent Kurdish state. Finally, in Kashmir, a variety of armed groups serve as instruments in a decades-old regional proxy war between Pakistan and India.

Sources of Insurgency

As suggested in the previous section, it is difficult to generalize about insurgency with any degree of confidence given the great differences in the motivations and capabilities of insurgent groups and governments, as well as the unique circumstances of each insurgency. It does seem reasonable, however, to conclude that insurgency is likely to remain a major feature of the international security environment. As one analyst has concluded, "[a]s long as there are people frustrated to the point of violence but too weak to challenge a regime in conventional military ways, insurgency will persist."[31] Although insurgents often engage in brutal and immoral behavior, such as the killing of noncombatants, insurgency can be a rational, low-cost strategy for pursuing political objectives. A variety of other factors are likely to prove conducive to existing conflicts and help generate new ones.

First among these is the problem of failed or failing states. Among other things, states with fragile, corrupt, or incompetent political institutions lack the resources and capacity to mediate violent or potentially violent intergroup conflicts.[32] Chester Crocker put it well when he concluded that as states fail,

---

[30] For more on the insurgency in Colombia, see Angel Rabasa and Peter Chalk, *Colombian Labyrinth: The Synergy of Drugs and Insurgency and its Implications for Regional Stability*, Santa Monica, Calif.: RAND Corporation, MG-1339-AF, 2001; Brian Michael Jenkins, "Colombia: Crossing a Dangerous Threshold," *The National Interest,* Winter 2000/2001; and Joaquin Villalobos, "Why the FARC is Losing," *Semana* (Bogata), Foreign Broadcast Information Service, July 14, 2003.

[31] Steven Metz, *The Future of Insurgency*, Carlisle Barracks, Pa.: U.S. Army Strategic Studies Institute, December 10, 1993, p. 1.

[32] Nick Donovan, Malcolm Smart, Magui Moreno-Torres, Jan Ole Kiso, and George Zacharaiah, "Countries at Risk of Instability: Risk Factors and Dynamics of Instability," back-

> the balance of power shifts ominously against ordinary civilians and in favor of armed entities operating outside the law. . . . [T]hey find space to operate in the vacuums left by a declining or transitional state—and they eat what they kill.[33]

The withdrawal of the state's presence—assuming of course that it had any to begin with—creates a vacuum that insurgents can use to create sanctuary, generate resources, and carry out training. It goes without saying that these lawless "gray areas" are essential to the movement, since if the state did have a real presence, such activities would necessarily be extremely difficult if not impossible.[34]

Until recently, most of these ungoverned spaces were in rural areas. Indeed, even today, some 20 countries—physically remote, low-density hinterlands—serve as havens for insurgent, terrorist, and criminal organizations—that is, "armed groups" that are challenging the state's monopoly on the use of force.[35] However, this rural locus for insurgents is gradually being eclipsed by urban environments that are increasingly conducive to rebel activity, as demonstrated dramatically in Iraq.[36] The explanation for this development is relatively simple. Worldwide, nearly all population growth will take place within the developing world's cities.[37] Over the next 20 years, nearly 4 billion of the world's expected population of 10 billion people will live in these urban areas.[38]

---

ground paper, London, UK: Prime Minister's Strategy Unit, February 2005.

[33] Chester A. Crocker, "Bridges, Bombs, or Bluster?" *Foreign Affairs*, September–October 2003, p. 32.

[34] For a discussion of such gray areas, see Max G. Manwaring, ed., *Gray Area Phenomena: Confronting the New World Disorder*, Boulder, Colo.: Westview Press, 2003.

[35] Shultz, Farah, Lockard (2004, p. 3).

[36] Jennifer Morrison Taw and Bruce Hoffman, *The Urbanization of Insurgency: The Potential Challenge to U.S. Army Operations*, Santa Monica, Calif.: RAND Corporation, MR-398-A, 1994.

[37] Population Information Program, The Johns Hopkins Bloomberg School of Public Health, "Meeting the Urban Challenge," *Population Reports*, Vol. XXX, No. 4, Fall 2002.

[38] Mike Davis, "Planet of Slums: Urban Involution and the Informal Proletariat," *New Left Review*, No. 26, March–April 2004.

Insurgents must go where the people and resources are, and as these migrate into cities, the insurgents must follow them.[39] Increasingly, cities, not the rural areas of the classic Maoist insurgents, are the only real sources of power. Large numbers of unemployed, bored, and restless young men serve as a recruitment reservoir. Densely populated with large numbers of disgruntled residents, these loosely or barely governed "feral" cities can serve as echo chambers for popular grievances far more readily than rural areas can.[40] Awash in weapons, sprawling, and often ringed by unmapped shantytowns, such settings are proving increasingly attractive to insurgents, who are able to live and operate free from the scrutiny of the security forces, which are often outgunned, unwilling, or unable to even enter these intimidating environments.[41]

In addition to state weakness and failure, a variety of other factors may help sustain existing insurgencies and contribute to new ones:

- environmental decay, the failure of economic development to meet popular expectations, and mass discontent arising from the process of globalization[42]
- "ethnic affiliations, intense religious convictions, and youth bulges"[43]
- the widespread availability of cheap and deadly small arms
- the presence of alluvial diamonds, gemstones, old-growth timber, and other precious commodities, which serve both as a motivat-

---

[39] Max G. Manwaring, *Shadows of Things Past and Images of the Future: Lessons for the Insurgencies in Our Midst*, Carlisle, Pa.: U.S. Army War College, Strategic Studies Institute, November 2004, p. 36.

[40] The phrase "feral cities" is taken from Richard J. Norton, "Feral Cities," *Naval War College Review*, Vol. LVI, No. 4, Autumn 2003.

[41] Beckett (2003, pp. 237–238).

[42] Metz and Millen (2004, p. 1).

[43] National Intelligence Council, *Mapping the Global Future: Report of the National Intelligence Council's 2020 Project*, Washington, D.C., December 2004, p. 97.

> ing factor for internal conflicts and as means for sustaining an armed struggle.[44]

But as important as these factors no doubt are, they do not give the full picture. As with other political phenomena, monocausal explanations of insurgency are inadequate. The presence of grievances is hardly sufficient, since as Walter Laqueur has noted, grievances are part of every human society. Insurgents succeeded in Cuba but failed elsewhere in the hemisphere, "despite the fact that Cubans had less objective reason to feel aggrieved than many other Latin Americans."[45] Although writing about revolutions, Forrest D. Colburn's conclusions are equally applicable to insurgency:

> There are no sufficient causes. The [20th] century's greatest revolutionary stressed the contingency of the Russian Revolution . . . Revolution defies not just established political and economic realities, but also the petty calculations of interest and advantage that comprise so much of everyday life. The prosaic is set aside.[46]

This last point is particularly relevant. The life of political violence has exercised a romantic pull on people ranging from the prosperous American supporters of the Provisional Irish Republican Army, through the university professors who joined Sendero Luminoso in Peru, to the middle-class British Muslims who attacked the London Underground in July 2005.[47] Writing during the mid-1960s, as the "wind of revolu-

---

[44] Philippe Le Billon, *Fuelling War: Natural Resources and Armed Conflict*, Oxford: Routledge, Adelphi Paper No. 373, March 2005, p. 9, and Nicolas Cook, "Diamonds and Conflict: Background, Policy, and Legislation," Washington, D.C.: Congressional Research Service, July 16, 2003, pp. 2–3. For a discussion of the Khmers Rouges' exploitation of gems and hardwoods, see Nate Thayer, "Rubies Are Red," *Far Eastern Economic Review*, February 7, 1991, p. 30.

[45] Walter Laqueur, *Guerrilla: A Historical and Critical Study*, Boston, Mass.: Little, Brown and Company, 1976, p. 380.

[46] Forest D. Colburn, *The Vogue of Revolution in Poor Countries*, Princeton, N.J.: Princeton University Press, 1994, p. 105.

[47] Anne Applebaum, "The Discreet Charm of the Terrorist Cause," *Washington Post*, August 3, 2005, p. A19.

tion" was in full force in the postcolonial countries of the developing world, Robert Taber observed that "the will to revolt . . . seems to be something more than a reaction to political circumstances."[48]

From the 1950s through the 1980s, the urge to rebel was expressed primarily, although not exclusively, through the vehicle of Marxism-Leninism. With communism in tatters, the most violent opposition to the current international order—an order dominated by the United States and its prosperous liberal democratic cohorts—is now found among the most extreme fringes of the Muslim world.

Insurgency and U.S. Security

During the Cold War, most senior policymakers believed they had little discretion with respect to countering insurgencies. With the entire world viewed as an arena for superpower competition, virtually any even remotely pro-Western state facing an insurgency was likely to receive at least some U.S. assistance. The list of such recipients is long. In addition to the obvious, such as South Vietnam and El Salvador, the United States supported counterinsurgency in Thailand, the Philippines, Greece, Bolivia, Peru, Uruguay, and Ethiopia, to name but a few.

But ever since the collapse of the Soviet Union, insurgency has been stripped of its broader geopolitical context, at least for U.S. national security planners. Absent a Soviet involvement, insurgencies during the past 15 years have rarely been seen as particularly threatening to U.S. interests. There have been important exceptions, of course, such as Colombia, where since the 1980s the United States has supported the struggle against the leftist Fuerzas Armadas Revolucionarias de Colombia (Revolutionary Armed Forces of Colombia [FARC]), albeit primarily because of FARC's involvement in narcotics trafficking. And today, insurgencies in Iraq, and to a much lesser extent Afghanistan, are high on the U.S. government's national security agenda. Indeed, defeating

---

[48] Robert Taber, *The War of the Flea: Guerrilla Warfare in Theory and Practice*, London, UK: Paladin, 1974, p. 18.

the insurgency in the former is deemed a vital national security interest of the United States, according to senior U.S. officials.[49]

That said, relatively little policy exists to guide officials as they work to improve the U.S. government's understanding of, and response to, current and emerging insurgent challenges. For example, the most recent edition of the *National Security Strategy of the United States,* published in March 2006, makes no explicit reference to insurgent threats, with the exception of brief mentions of Iraq, Nepal, and insurgency as a type of irregular challenge.[50] The subject of insurgency receives little more attention in other U.S. government policy statements. According to the *National Defense Strategy of the United States* (2005), adversaries use "irregular methods," such as terrorism and insurgency, to "erode U.S. influence, patience, and political will" and, in so doing, "compel strategic retreat from a key region or a course of action."[51] "Illegal armed groups," according to the *National Military Strategy of the United States* (2004), "menace stability and security," and "[e]ven some individuals may have the means and the will to disrupt international order."[52]

These observations are certainly correct. But as the U.S. armed forces and other national security organizations grapple with the challenge of preventing and responding to insurgency, it may be worthwhile to consider two additional aspects of the broader context in which insurgencies are taking place. The first involves state sponsorship. By all accounts, states today are less involved in supporting "illegal armed groups" than they were during the Cold War, although, as suggested above, the Soviet Union and its allies appear to have given less aid than

---

[49] See, for example, Zalmay Khalilzad, U.S. ambassador-designate to Iraq, "Priorities for U.S. Policy in Iraq," statement submitted to the Senate Foreign Relations Committee, Washington, D.C., June 7, 2005.

[50] White House, *National Security Strategy of the United States of America*, Washington, D.C., September 2006, pp. 18, 20, and 49.

[51] U.S. Department of Defense, *The National Defense Strategy of the United States of America*, Washington, D.C., March 2005, p. 3.

[52] Joint Chiefs of Staff, *National Military Strategy of the United States of America*, Washington, D.C.: U.S. Department of Defense, 2004, pp. 4–5.

was commonly assumed at the time.[53] While "self-financing," through linkages with the drug trade, extortion, and trafficking in precious commodities, was a feature of some Cold War insurgencies, it appears to be more widespread and more important today. Similarly, nonstate support from diaspora communities and ideological "fellow travelers," while evident during the Cold War, has assumed much greater importance in the current security environment.[54]

That said, some governments do continue to view insurgent groups as instruments of statecraft. The Pakistani government, for example, has given intelligence, training, weapons, and other resources to Kashmiri separatists as a relatively low-cost, low-risk way to keep pressure on India.[55] In the future, more states may find the use of this indirect form of warfare attractive, thereby making insurgency a more-prominent feature in the conflict among nations.[56] State sponsorship of terrorism has long been a concern for the United States, which employs a variety of sanctions and other instruments (including, on occasion, military force) to punish governments that assist terrorists. In the future, if U.S. policymakers deem the stakes to be sufficiently high, they may consider the use of combat power to deter or punish a state that supports an insurgency. U.S. counterinsurgency may thus in some cases involve more than supporting a host nation or using American military forces against insurgents; it may include operations against a third country.

---

[53] S. Neil McFarlane, "Successes and Failures in Soviet Policy Toward Marxist Revolutions in the Third World, 1917–1985," in Mark N. Katz, ed., *The USSR and Marxist Revolutions in the Third World*, Cambridge, UK: Woodrow Wilson International Center for Scholars and Cambridge University Press, 1990, p. 30.

[54] For a discussion of state and nonstate assistance, see Daniel L. Byman, Peter Chalk, Bruce Hoffman, William Rosenau, and David Brannan, *Trends in Outside Support for Insurgent Movements*, Santa Monica, Calif.: RAND Corporation, MR-1405-OTI, 2001. See also Daniel Byman, *Deadly Connections: States that Sponsor Terrorism*, Cambridge, UK: Cambridge University Press, 2005a.

[55] See U.S. Department of State, online report, 2000, and Council on Foreign Relations, "Terrorism Havens: Pakistan," Web page, updated December 2005. The Pakistani government denies giving such aid.

[56] Metz (1993, p. 25).

Second, national security planners will have to come to grips with the challenges created by what some government officials, analysts, and journalists are calling the "global jihadist insurgency."[57] This concept has taken a variety of forms, the most prominent of which is the notion that it has evolved from a terrorist organization into a worldwide movement that includes autonomous terrorist cells, as well as guerrilla armies in places, such as the Philippines, Kashmir, Chechnya, Pakistan, Afghanistan, and now Iraq. For some experts, this global movement also includes (at least potentially) non-Islamic elements, engaged for common operational or even ideological purposes.[58]

As Jason Burke has noted, two main forms of violent Islamist militancy emerged from the crucible of the anti-Soviet war in Afghanistan:

> local insurgencies which aimed to impose Islamic law on specific countries, and transnational or international militancy, which targeted an ill-defined range or people deemed responsible for the ills of the umma, the global Muslim community.[59]

Osama bin Laden and his circle have attempted to transform these local struggles and "internationalize" them as part of a global campaign against the United States, "crusaders," Zionists, and other perceived enemies of the *umma*. Militants have been urged to abandon their struggles to create an Islamic state within their own countries and redirect their violence (both physical and rhetorical) against Israel, the United States, and non-Muslims, the wellsprings of a supposed global crusade against Islam. While some Muslim insurgents, such as the Moro Islamic Liberation Front in the Philippines, have rejected this

---

[57] See, for example, Paul Rich, "Al Qaeda and the Radical Islamic Challenge to Western Strategy," *Small Wars & Insurgencies*, Vol. 14, No. 1, 2003, p. 46.

[58] For an assessment of this issue, as well as a discussion of the possible convergences among contemporary terrorists, insurgents, and criminals, see Angel Rabasa, Peter Chalk, R. Kim Cragin, Sara A. Daly, Heather S. Gregg, Theodore W. Karasik, Kevin A. O'Brien, and William Rosenau, *Beyond Al Qaeda, Part 2: The Outer Rings of the Terrorist Universe*, Santa Monica, Calif.: RAND Corporation, MG-430-AF, forthcoming.

[59] Jason Burke, "Special Report: Al Qaeda After Spain," *Prospect* (London), May 27, 2004.

globalist stance, internationalization has occurred elsewhere. European jihadists, for example, are not seeking to establish an Islamic state on the continent. The objectives of jihadists in Iraq are less clear. Many jihadists who make their way to Iraq do so simply to fight against the United States and to acquire skills that will be useful in other armed struggles.[60] Others may have more-ambitious goals. For example, the July 9, 2005, letter between al Qaeda leaders Ayman al-Zawahiri and Abu Musab al-Zarqawi describes a second stage in the conflict after the Americans have departed. The key goal in that second stage is to "[e]stablish an Islamic authority or amirate, then develop it and support it until it achieves the level of a caliphate—over as much territory as you can to spread its power in Iraq."[61]

How should these developments shape U.S. responses? Given the scope and nature of the challenge, it was sensible for the Bush administration to move beyond its narrow conception of a "global war on terrorism" and accept that a broader campaign was required.[62] Senior policymakers, such as Secretary of Defense Donald Rumsfeld, have mentioned the emergence of a global insurgency,[63] and in the judgment of the Bush administration, Iraq is its most important manifestation.

But countering a global insurgency is a formidable challenge for the United States, and as policymakers craft a counterstrategy, they should keep in mind the inherent difficulties of such an approach. Global insurgency is not a new problem. During the Cold War, the

---

[60] Olivier Roy, "Britain: Homegrown Terror," *Le Monde Diplomatique*, August 2005, p. 5.

[61] Some scholars question the authenticity of this letter. See Juan Cole, "Zawahiri Letter to Zarqawi: A Shiite Forgery?" *Informed Comment: Thoughts on the Middle East, History, and Religion*, Web site, October 14, 2005. The full text of the letter is available on the Web, in Office of the Director of National Intelligence, News Release, February 2005.

[62] In July 2005, there were media reports that the administration was going to replace "global war on terrorism" with the phrase "struggle against violent extremism." Although the latter phrase has been used in some speeches (e.g., by Chairman of the Joint Chiefs Richard Meyer), no official documents have been published announcing the change. See Center for Media & Democracy, *Global Struggle Against Violent Extremism*, Web site, last updated February 13, 2006.

[63] See for example Secretary of Defense Rumsfeld's speech, "U.S. Refocusing Military Strategy for War on Terror, Rumsfeld Says," remarks delivered at United States Military Academy Commencement, Michie Stadium, West Point, N.Y., May 29, 2004.

United States waged a worldwide campaign against a loosely structured international movement that employed terrorism, insurgency, and subversion against U.S. friends and allies across the global South. If the United States is to wage a worldwide counterinsurgency campaign, policymakers must continually remind themselves that actions in one country or region are likely to have a ripple effect elsewhere and that no measures should be taken without first considering how actors and audiences elsewhere might receive them. More important, it will be essential to avoid a "cookie-cutter" approach, an all-too-familiar pitfall from the Cold War era, when decisionmakers developed universalistic responses to communist revolution that misguidedly disregarded local contingencies. Finally, national security planners must be careful in their use of the term *global insurgency*. Sloppy usage could hand bin Laden and other opponents of the United States a propaganda victory by seeming to suggest that Western countermeasures are in reality a worldwide struggle against the *umma*, as the jihadists have long claimed.

Conclusion

As the eminent British strategist B. H. Liddell Hart noted in 1967, as the United States was embroiled in a campaign against the National Liberation Front in South Vietnam, the

> problems of guerrilla warfare are of very long standing, yet manifestly far from understood—especially in those countries where everything that can be called "guerrilla warfare" has become a new military fashion or craze.[64]

In many respects, the United States today, as during early periods of heightened official interest, is approaching the challenge as if it were entirely new. Although what U.S. officials are terming the "global insurgency" is clearly the first of its kind, insurgency is a longstanding feature of the international security environment. Given the enduring

---

[64] B. H. Liddell Hart, *Strategy*, 2nd rev. ed., New York: Meridian, 1991, p. xv.

factors that help create and sustain these movements and the strategy's attraction for individuals, groups, and states that are seeking to change the existing political order, it seems likely that insurgency will remain a prominent element in world politics. If that is the case, counterinsurgency must evolve beyond a mere "fashion or craze" and become an established part of U.S. national security policy.

The next chapter continues our discussion of the insurgency phenomenon, moving from an exploration of trends to lessons learned from counterinsurgency operations over the last century. Specifically, the chapter identifies four broad principles to guide national security planners as they seek to devise strategies to defeat insurgencies.

CHAPTER THREE

The Challenge of Counterinsurgency: Lessons from the Cold War and After

Introduction

For the first time since the end of the Cold War, DoD has identified irregular challenges as one of the four major threats to U.S. interests.[1] Although insurgencies are only one of the irregular challenges identified at the DoD level, both the Army and Marines are taking them seriously; the Army published new doctrine in 2004, and the Marines are rewriting their famous manual on small wars," which dates back to 1940.[2] This marks the third time in post–World War II history that the nation's defense establishment has undertaken a significant effort to understand and respond to the threats posed by what has been referred to at various times as "indirect aggression," "low-intensity conflict," "remote area conflict," "irregular warfare," and insurgency.

The first "counterinsurgency era" began during the administration of President John F. Kennedy with a flurry of initiatives designed to shore up friendly governments in the developing world threatened by communist-inspired insurgency and subversion.[3] America's defeat

---

[1] The other threats are *traditional*, *catastrophic*, and *disruptive*. See Joint Chiefs of Staff (2004).

[2] See U.S. Army (2004) and U.S. Marine Corps, *Small Wars Manual*, Manhattan, Kan.: Sunflower University Press, 2004. (Originally published by the U.S. Marine Corps in 1940.)

[3] The phrase "counterinsurgency era" is from Douglas S. Blaufarb, *The Counterinsurgency Era: U.S. Doctrine and Performance, 1950 to the Present,* New York: The Free Press, 1977. For more on the Kennedy administration's counterinsurgency fervor, see William Rosenau,

in Southeast Asia brought that era to a close and ushered in a period in which the military refocused its attention on what it understood to be its central organizing principle: preparing to fight and win large-scale conventional war against "symmetrical" adversaries.[4]

In the early 1980s, a major insurgency in El Salvador, backed by the Soviet Union and Cuba, led to a second counterinsurgency era. Like Kennedy, President Ronald Reagan was convinced that the United States was dangerously unprepared for waging what was then termed "low-intensity conflict," and his administration took a series of steps to improve U.S. capabilities, such as the creation of U.S. Special Operations Command (SOCOM) in 1987.[5] DoD's enthusiasm for low-intensity conflict proved to be short-lived, however.[6] The collapse of the Soviet Union in 1991 convinced most policymakers that counterinsurgency, unconventional warfare, psychological operations, and other "nontraditional missions" were at best irrelevant and at worst costly anachronisms in the emerging security environment.

Today we are witnessing the dawn of a third counterinsurgency era. The new attention to the problem is primarily a response to the conflicts in Afghanistan and Iraq, in which irregular adversary forces, displaying flexibility, lethality, resilience, and operational depth, have frustrated the world's most capable armed forces. The reconceptualization of the global war on terrorism as a struggle against a broader set of

---

"The Kennedy Administration, U.S. Foreign Internal Security Assistance and the Challenge of 'Subterranean War,' 1961–63," *Small Wars & Insurgencies*, Vol. 14, No. 3, Autumn 2003, pp. 65–99.

[4] U.S. nuclear forces received renewed investment and attention during this period as well. Deterring, and even preparing to fight, a nuclear war was a high-priority mission for DoD, but it was the central organizing principle only for the Strategic Air Command and other specialized forces.

[5] For more on the efforts to reinvigorate U.S. special operations forces during this period, see Susan Marquis, *Unconventional Warfare: Rebuilding U.S. Special Operations Forces*, Washington, D.C.: Brookings Institution Press, 1997.

[6] A 1990 RAND report for the Army detailed steps the Army could take to improve its counterinsurgency capabilities, but without a broader consensus recognizing the importance of counterinsurgency, little progress was made. See Stephen T. Hosmer, *The Army's Role in Counterinsurgency and Insurgency*, Santa Monica, Calif.: RAND Corporation, R-3947-A, 1990.

transnational Islamist actors that includes insurgents has contributed further to the belief that the United States must enhance its ability to counter insurgent threats.[7] As in earlier counterinsurgency eras, Army Special Forces are being expanded; new doctrine is being developed; and the professional military literature is awash with articles that assess earlier counterinsurgency campaigns, consider new operational concepts, and examine organizational and institutional implications for the armed forces.[8]

Although history suggests that official interest in counterinsurgency is often fleeting, it may also be true that the United States is facing a generations-long struggle against a globally distributed jihadist adversary for whom insurgency is an important political-military tool. As U.S. Army Chief of Staff General Peter Schoomaker said in 2004, this conflict "is a little bit like having cancer. You may get in remission, but it's never going to go away in our lifetime."[9] If this claim is correct, the new counterinsurgency era is likely to be protracted, and substantial resources, both human and material, will have to be committed to preparing the military to respond. Allocating these resources effectively requires an understanding of the nature of the challenges counterinsurgency poses, whether the United States allocates them directly or indirectly through support for friendly governments threatened by insurgents.

---

[7] See, for example, Linda Robinson, "Plan of Attack," *U.S. News & World Report*, August 1, 2005, p. 26.

[8] For more on the recent expansion of Special Forces and other SOF, see Michael Duffy, Mark Thompson, and Michael Weisskopf, "Secret Armies of the Night," *Time*, June 23, 2003; Andrew Feickert, "U.S. Special Operations Forces (SOF): Background and Issues for Congress," Congressional Research Service, June 9, 2005; and U.S. Special Operations Command, *2005 Annual Report*, 2005, p. 18. New military doctrine includes U.S. Air Force, *Foreign Internal Defense*, Maxwell Air Force Base, Ala.: Air Force Doctrine Center, Doctrine Document 2-3.1, May 10, 2004, and U.S. Army (2004). Recent examples in the professional military literature include Manwaring (2004); William Brian Downs, "Unconventional Airpower," *Air & Space Power Journal*, Spring 2005; Robert R. Tomes, "Relearning Counterinsurgency Warfare," *Parameters*, Vol. XXXIV, No. 1, Spring 2004; and Lester Grau, "Guerrillas, Terrorists, and Intelligence Analysts," *Military Review*, July–August 2004.

[9] Quoted in Robert Burns, "Army Chief Says Islamic Extremist Threat is Like a 'Cancer' that Will Linger," *Associated Press*, June 15, 2004.

Toward that end, this chapter will offer four very broad principles national security planners should bear in mind as they develop strategies, forces, and technology for counterinsurgency. These tenets, derived from an analysis of counterinsurgency "classics," as well as more-recent studies, are impossible to consider in isolation from each other, and so, inevitably, there is overlap in the discussion. Neither are these four propositions intended to be all-inclusive.

Before discussing these propositions, we raise several caveats. First, given the consensus among analysts, scholars, and practitioners that early intervention is far cheaper and much more effective than combating a mature insurgency, the emphasis in this chapter is on responding to "incipient" insurgency, defined by one scholar as "the period beginning with the first discussions and small meetings about insurgency, through its earliest organizational phase, to the outbreak of guerrilla war."[10] Second, it is assumed that the reader is already familiar with the cardinal propositions of counterinsurgency: that military power alone is incapable of securing the defeat of an insurgency and that success or failure will always be decided in the political sphere. However, the primacy of the political is often forgotten in the heat of a counterinsurgency campaign. While this chapter considers the role of combat power, the stress is on other aspects of the counterinsurgency challenge. Third, it is also assumed that the reader understands that, in counterinsurgency as in all things, persistence and perseverance are essential, as is a sound strategy. Fourth, much of the literature on the subject stresses the importance of proper organization, such as establishing a single "supremo" responsible for all military and civilian aspects of the campaign. But as desirable as rational and effective organizational arrangements are in theory, they have almost never been realized in practice. The British experience during the Malayan Emergency (1948–1962) is the canonical case of successful organization, but analysts are hard pressed to find another comparable example. American strategists have devoted vast energy to grappling with administrative, organizational, and command-and-control arrangements, but in the final analysis, such aspects are usually secondary or tertiary factors

---

[10] George K. Tanham, "Indicators of Incipient Insurgency," unpublished paper, 1988, p. 1.

with respect to the conflict's outcome.[11] Thus, they are not considered in this chapter.

Finally, while this chapter does examine a number of counterinsurgency tactics, such as the use of amnesty, the focus is on broader operational and strategic principles. Among other things, this focus will help avoid an overemphasis on technique, a frequent pitfall in discussions of counterinsurgency, which often become bogged down in technical aspects of the challenge, such as the ratio of government troops to insurgents required to prevail in counterinsurgency.

Before considering the four principles, it is important to provide additional context by considering several cross-cutting themes. It should be recognized that most insurgencies fail, their names (if they had them at all) lost to history. By definition, states have enormous resources relative to even the most robust of insurgencies. Life in the revolutionary underground is stressful, treacherous, and most often ends in death or imprisonment. At the same time, insurgents possess a number of significant advantages. First among these is information—quite simply, they know far more about their state adversary than the government knows about them, particularly in the insurgency's early phases. Given that the state is often oblivious to the insurgent's underground activities, the insurgents have a head start on the security forces and can exploit this strategic initiative.[12] As the insurgency matures, the state must commit resources to territorial security and the static defense of key facilities. Except in the most mature stages of the insurgency, when the movement has established "liberated zones," the insurgents have no territory to defend. Insurgents are poor, but as one scholar has observed, "the guerrilla has the freedom of his poverty."[13]

---

[11] See for example R. W. Komer, *The Malayan Emergency in Retrospect: Organization of a Successful Counterinsurgency*, Santa Monica, Calif.: RAND Corporation, R-0957-ARPA, 1972, particularly Chapter 3.

[12] Metz and Millen (2004, p. 23).

[13] Taber (1974, p. 22).

Four Principles

1. Understand the Adversary

As in any armed struggle, success in counterinsurgency requires a full appreciation of the adversary's strengths, weaknesses, and goals. In counterinsurgency, however, the opposition is likely to be far more elusive and opaque than it is in conventional conflicts between nation-states. Insurgencies, particularly in their formative stages, are by nature subterranean movements whose members "hide in plain sight" by living, training, and operating among a wider civilian population. With no uniforms to identify them and with much effort expended to remain clandestine, the cadres, leaders, and support members of an insurgency are unlikely to present obvious targets for intelligence and surveillance.

Indeed, if the state or occupying force were able to "see" the insurgents, there would likely be no insurgency in the first place, since a complete picture of the movement in its latent stage would give the incumbent most of what it needed to destroy the underground movement. Totalitarian regimes never have to confront insurgencies, at least within their own borders, since by definition these regimes have in place surveillance and control mechanisms that make the development of systematic rebellion impossible. Under totalitarianism, there is thus no "space" for an insurgency to germinate, and the actions of such regimes are not constrained by concern about domestic or international public opinion.[14]

Understanding the scope, capabilities, and intentions of a nascent insurgency requires a police and intelligence system capable of collecting information, analyzing it rigorously, and using the results to establish priorities for additional collection—the familiar "intelligence cycle" that all modern intelligence services employ. Human sources, who are able to provide insights into key issues, such as the movement's strategy, structure, and recruitment patterns, are likely to prove particularly important. As with other aspects of counterinsurgency, early action is critical. Gathering intelligence is easier in a movement's early phase,

---

[14] Laqueur (1976, p. 390).

before battle lines have been drawn and when the public is generally more willing to talk.[15] More important, early collection and analysis, properly utilized, can help a threatened government avoid strategic and operational errors and devise policies that will have the greatest chance of thwarting the insurgency in its early manifestations.

Typically, however, threatened regimes lack the capacity to collect, analyze, and act on intelligence information. As we will discuss in greater detail below, such governments are usually weak and corrupt, and in the case of intelligence, as with every other aspect of the state, "underadministration" is a chronic problem. Security services, including the police, are likely to have little presence within the disaffected minority communities that can play a key role in the conflict.[16] Police and intelligence agencies in the developing world are typically as unprofessional, underequipped, and undermanned as is the rest of the state's administrative apparatus.[17] These weaknesses make the institutions of the state particularly ripe for infiltration, a key insurgent tactic. Today in Iraq, the coalition has belatedly recognized that the insurgents—like earlier insurgents in South Vietnam, El Salvador, and Peru, to name just a few instances—have been able to gather invaluable intelligence on internal security operations by penetrating the country's police services.[18] Finally, U.S. support to threatened regimes has typically emphasized military equipment and training, rather than assistance to police and paramilitary forces.[19]

---

[15] Frank Kitson, *Low Intensity Operations: Subversion, Insurgency, and Peacekeeping*, London, UK: Faber and Faber Ltd., 1971, pp. 91–92.

[16] William Rosenau, " Recruitment Trends in Kenya and Tanzania," *Studies in Conflict and Terrorism*, Vol. 28, No.1, January–February 2005b, p. 7.

[17] George W. Allen, "Intelligence in Small Wars," *Studies in Intelligence*, Winter 1991, p. 23.

[18] U.S. Department of State and U.S. Department of Defense, Inspectors General, *Interagency Assessment of Iraq Police Training*, July 15, 2005, and John J. Lumpkin, "Insurgents Said to Be Infiltrating Security Forces," *Miami Herald*, October 22, 2004.

[19] This is dramatically evident today in Iraq but was a characteristic of U.S. counterinsurgency policy throughout the Cold War. For more on this point, see William Rosenau, *U.S. Internal Security Assistance to South Vietnam: Insurgency, Subversion, and Public Order*, New York: Routledge, 2005a, particularly Chapter 7.

The problem, however, extends beyond state capacity. There is also the problem of will. As one leading authority has observed, "[g]overnments that make an honest effort to know themselves and their enemies are in a better position to identify threats and to contemplate effective responses."[20] However, governments threatened by insurgency are typically reluctant to "face facts" and to come to terms with the true nature of the threats they face. As a consequence, countermeasures are applied belatedly and are likely to be more costly and less effective than if they had been employed earlier. Among other things, accepting that an insurgency is developing is a *de facto* admission of state failure. Insurgents, to the extent they are acknowledged at all, are often dismissed initially as mere "bandits," "foreign agitators," or "terrorists." With such a mindset, it is hardly surprising that many states are unwilling to commit the appropriate intelligence resources to developing a rigorous and comprehensive threat assessment.

Occupying forces, by definition "outsiders," face a similar and perhaps more formidable challenge, since they are less likely than incumbent governments to have the resident skills to assess an insurgent adversary, at least in the near term. Bluntly stated, most conventional armed forces lack the language skills and cultural awareness—so-called "social intelligence"[21]—necessary to develop the human sources required to penetrate and understand insurgent movements. Understanding the social structures from which insurgents emerge and through which they operate—for example, tribes and clans—requires what one analyst has termed "anthropological finesse."[22] What is more, the structure of some societies can negate Western strengths in terms of intelligence collection technology. This has been demonstrated dramatically in Iraq:

---

[20] Bard O'Neill, *Insurgency and Terrorism: From Revolution to Apocalypse,* Dulles, Va.: Potomac Books, 2005, p. 189.

[21] For an overview of "social intelligence" and the U.S. armed forces, see Megan Scully, "'Social Intel': New Tool for U.S. Military," *Defense News*, April 26, 2004.

[22] Montgomery McFate, "The Military Utility of Understanding Adversary Culture," *Joint Forces Quarterly*, No. 38, 3rd Qtr. 2005, p. 22.

> Penetrating Arab society, let alone an underground movement, is extremely difficult for Western military forces. Apart from the obvious language and physical barriers, the great strengths of Arab culture are based on family familiarity with the locale (strangers stand out immediately) and a preference for one-to-one/word of mouth communication. These three facets preclude the vast majority of methods (especially those based on SIGINT [signals intelligence]) to gain intelligence quickly by outside forces.[23]

Given the intelligence weaknesses and gaps discussed above, developing and effectively using an early warning system presents clear challenges. However, as a first step, governments concerned about the possible outbreak of insurgency can examine the indicators that U.S. civilian and military experts developed during and after the Cold War. The indicators listed here are not intended to be comprehensive. Rather, they are offered merely to illustrate the range of potentially relevant "observables":

- indicators of incipient rural insurgency[24]
  - absence of bright young people from a village
  - introduction of new words and phrases
  - hostile radio broadcasts
  - sudden change in religious beliefs
  - sudden change in customs
  - refusal of villagers to talk
  - officials leaving their posts
  - demonstrations and riots
  - presence of front groups

---

[23] Alistair Finlan, "Trapped in the Dead Ground: U.S. Counter-Insurgency Strategy in Iraq," *Small Wars and Insurgencies*, Vol. 16, No. 1, March 2005, p. 14. In counterinsurgency, military intelligence activities associated with conventional conflict, such as intelligence preparation of the battlefield and order of battle, are likely to be irrelevant or useful only if utilized in very different ways. In the end, intelligence techniques associated with police counterdrug and countergang operations, such as network analysis, are likely to prove more useful (Grau, 2004, p. 43).

[24] Tanham (1988).

- small raids for money and arms
- assassination of government representatives.

- enemy activity indicators—indigenous population[25]
  - unusual gatherings among the population
  - disruption of normal social patterns
  - increase in size of embassy or consulate staffs from a country or countries that support indigenous disaffected groups
  - reports of opposition or disaffected indigenous population receiving military training in foreign countries
  - infiltration of student organizations by known agitators
  - reports of payment to locals for engaging in subversive or hostile activities
  - evidence of the participation of paid and armed demonstrators in riots
  - refusal of population to pay or unusual difficulty to collect rent, taxes, or loan payments.

The indicators in the first bullet apply to rural insurgencies; it would be valuable to develop a similar set of indicators for insurgencies (e.g., Iraq in 2006) that have a strong urban component: Taken in isolation, none of these is particularly noteworthy, and some in fact may appear quite benign, but when considered in the context of local political conditions, they can suggest an analytical pattern.[26] A rich appreciation of the local political, social, and economic environment of course makes these indicators all the more functional. Finally, it should be noted that the usefulness of these tools is also a function of the robustness of the counterinsurgents' intelligence gathering and processing capabilities, which, as suggested earlier, may be quite modest.

A number of assessment tools are also available to measure progress against an insurgency in its more-mature phases. The following are some sample "campaign metrics":

---

[25] U.S. Army (2004, pp. E-1, E-2).

[26] [British] Army (2001, p. B-4-3).

- decline in the number of successful assassinations of government officials, religious figures, and business leaders
- rise in insurgent leader casualties and defections
- willingness of population to provide useful or "actionable" intelligence
- "market metrics": payments to individuals to carry out terrorist attacks suggest the insurgency failing to build or maintain popular support
- independence and effectiveness of host-nation security forces.[27]

2. Build State Capacity and Presence

As mentioned in the preceding section, the lack of capacity in police and intelligence is a critical shortfall for many governments facing insurgencies. That lack of capacity contributes to the growth of insurgent movements in a number of ways. Corrupt, lazy, and incompetent policemen—or simply the shortage or absence of policemen of any caliber—signals to the general public a regime's inability or unwillingness to provide for public order, one of the fundamental responsibilities of any modern polity. The public-safety vacuum, in turn, allows the growth of private militias, criminal elements, and other violent nonstate actors and indicates to potential insurgents that the state may be a less formidable adversary than previously supposed. Finally, the absence of competent police—who ideally would have served as a regime's early warning system—allows the underground movement to develop unwatched and unmolested. In South Vietnam during the late 1950s and early 1960s, a public-safety vacuum in the countryside highlighted the profound weaknesses of the Diem government and contributed to the rapid spread of the Viet Cong.[28] Without a shield of public order in place, it is impossible for an embattled government to carry

---

[27] Andrew F. Krepinevich, "Are We Winning in Iraq?" testimony before U.S. House of Representatives, Committee on Armed Services, March 17, 2005; Richard Betts, "Stability in Iraq?" unclassified draft of paper prepared for the Center for the Study of Intelligence, Central Intelligence Agency, June 7, 2005.

[28] Rosenau (2005a, p. 37).

out the political, social, and economic steps necessary for thwarting an insurgency.

Unfortunately, however, the problem of the lack of capacity is rarely confined to civilian security forces. In countries facing a burgeoning insurgency, most if not all the institutions of the state—e.g., the civil service, health care, schools—are weak (and sometimes virtually nonexistent), riddled with corruption, and incapable of fulfilling their most basic responsibilities. More broadly, these states often lack the political and administrative capacity to govern. In short, nations facing insurgencies are typically "failing states," that is, they are characterized by "declining public order, rising domestic violence, stagnating economies, and infrastructure deteriorating because of the lack of basic maintenance."[29]

Developing state institutions and extending the government's presence are thus key requirements for effective counterinsurgency. In the judgment of Robert Thompson, a leading counterinsurgency theorist who played an instrumental role in suppressing the communist insurgency in Malaya,

> Without a reasonably efficient government machine, no programmes or projects, in the context of counter-insurgency will produce the desired results . . . The correction of these [administrative] weaknesses is as much part of counter-insurgency as any military operation.[30]

Extending the state's presence and capacity serves a number of interrelated purposes. Insurgencies are in part a struggle for the control of a given population. In that contest, evidence of resilience, strength, and what one writer has termed "[t]he appearance of being the eventual victor" serves as "an incalculable force multiplier."[31] Vigorous and

---

[29] William J. Olson, "The New World Disorder," in Max G. Manwaring, ed., *Gray Area Phenomena: Confronting the New World Disorder*, Boulder, Colo.: Westview Press, Inc., 1993, p. 11.

[30] Robert Thompson, *Defeating Communist Insurgency: Experiences from Malaya and Vietnam*, London, UK: Chatto & Windus, 1967, p. 51.

[31] Anthony James Joes, *Resisting Rebellion: The History and Politics of Counterinsurgency*, Lexington, Ky.: University of Kentucky Press, 2004, p. 235.

effective administration, the development of political institutions, road-building, and other government actions can help convince local populations that the state is determined to prevail. In the formative stages of an insurgency, most of the public are "fence-sitters" who are not committed to either the government or the opposition. State-building efforts can win converts by showing the public that the government can meet its most fundamental responsibilities and, in so doing, demonstrate that the state can deliver what the insurgents can only promise.[32]

Additionally, such measures can convince key international audiences that the threatened government has made a credible commitment to defeating the insurgency. This is particularly important in terms of garnering international assistance for the counterinsurgency campaign. During the 1980s, critics of U.S. support to counterinsurgency in El Salvador questioned whether a regime incapable of governing was worthy of survival.[33] Building state capacity and presence, and demonstrating an ability to govern, can shore up a threatened regime's credibility and standing, both domestically and abroad.

Establishing or strengthening democratic institutions can also strengthen the counterinsurgents' cause, both at home and abroad.[34] Giving its citizens legal opportunities to express grievances, organize political parties, run for office, and vote in honest elections demonstrates that the government is willing and capable of enacting essential reform and thereby undermines domestic and international sympathy for the insurgents.[35]

---

[32] D. Michael Shafer, *Deadly Paradigms: The Failure of U.S. Counterinsurgency Policy*, Princeton, N.J.: Princeton University Press, 1988, p. 117.

[33] Benjamin C. Schwarz, *American Counterinsurgency Doctrine and El Salvador: The Frustrations of Reform and the Illusions of Nation Building*, Santa Monica, Calif.: RAND Corporation, R-4042-USDP, 1991, p. 73.

[34] Joes (2004, p. 234).

[35] Whether democratic reforms are necessary to defeat insurgents is a matter of debate. They are, however, increasingly necessary in the court of world opinion. Given its new strategy of exporting freedom, it is unlikely that the United States will give support to regimes unwilling to institute such reforms.

Role of Good Government. Two other related factors can also be invaluable to states combating insurgencies. First, taking forceful measures to reduce corruption can have both an instrumental effect—by attacking the distortions, inefficiencies, and misallocations of resources created by graft—and a more-important political purpose: demonstrating the state's essential legitimacy. Second, conducting counterinsurgency within a legal framework is essential if the state's campaign is to have any semblance of legitimacy. Central elements in such a framework include the right to trial, due process, and an absolute ban on torture. Of course, from the insurgents' point of view, the question of legality is moot, since the rebels, in rising up against the incumbent power, are in effect denying that its adversaries have the right to make and enforce laws.

In one respect, a sound legal framework can serve as an enormous disadvantage, since insurgents are free to operate without the bonds that restrain government forces. However, the alternative of "lawless" counterinsurgency is likely to prove fatal to the threatened regime, since states that operate outside a legal framework will probably be unable to command popular support or international assistance. In the end, most observers are likely to accept Thompson's conclusion that "[a] government which does not act in accordance with the law forfeits the right to be called a government and cannot then expect its people to obey the law."[36]

Leverage and Reform. American conduct in South Vietnam during the 1960s and in Iraq today are obvious exceptions to a principle that the U.S. government has embraced since the early days of the Cold War, that the threatened "host nation" is ultimately responsible for the success or failure of the campaign against internal rebellion. "Entirely uncomfortable with the role of proconsul,"[37] the United States traditionally prefers an indirect approach that stresses working with and through local governments, a strategy one writer deemed "colonial-

---

[36] Thompson (1967, p. 52).

[37] Schwarz (1991, p. 39).

ism by ventriloquism."[38] Given U.S. political and cultural traditions, such an approach is understandable, but it comes with costs. Unlike a colonial power waging counterinsurgency in one of its far-flung possessions, the United States must continuously bargain with local leaders to induce them to adopt the proper measures to thwart the insurgency. This is particularly problematical in the case of the political, economic, and social reforms that form part of America's counterinsurgency repertoire. Nostrums, such as land reform, anticorruption initiatives, and the rule of law are easy to prescribe, but getting a government to enact them is another matter. Selfishly but often correctly, local elites perceive such reforms as threats to their own power and prestige. In enacting them, therefore, these elites run the risk of reforming themselves out of existence. In the face of a mature insurgency, the task becomes more daunting, as the United States is pressuring the embattled regime to engage in "self-reform in crisis."[39]

Finally, as has been widely noted in the literature, a heightened U.S. commitment to a threatened government can have the paradoxical effect of reducing U.S. leverage over the host nation. The failure of a client government to self-reform is often met with U.S. threats to cut assistance or withdraw entirely. However, host nations typically ignore such attempts at coercion, secure in the belief that national interest compelled U.S. involvement in the first place, and are therefore unlikely to permit any serious reduction in support from their American patrons.[40]

3. Control the Population

There is widespread agreement among theorists and practitioners that the population plays an essential role in any insurgency, with the people serving as an invaluable (if not irreplaceable) resource base, both mate-

---

[38] William Odom, *On Internal War: American and Soviet Approaches to Third World Clients and Insurgents*, Durham, N.C.: Duke University Press, 1992, p. 63.

[39] Blaufarb (1977, pp. 305–306).

[40] Schwarz (1991, p. 40) and Douglas J. Macdonald, *Adventures in Chaos: American Intervention for Reform in the Third World*, Cambridge, Mass.: Harvard University Press, 1992, Chapters 4 and 5.

rial (e.g., food, recruits, arms) and political. "Without the consent and active aid of the people," concludes one analyst, "the guerrilla would be merely a bandit, and could not long survive."[41] While this is certainly true for a mature insurgency, it is by no means certain that such vigorous support is necessary during the conflict's infancy when, as has been noted above, most of the population is uncommitted. It does seem clear, however, that an insurgency is unlikely to grow in an environment in which the population is actively hostile.[42]

Ensuring that the fence-sitters do not end up siding with the insurgency should obviously be a central element of any sound counterinsurgency strategy. There are different schools of thought on how best to do that.

One school argues that both the populace and insurgents are rational actors who will respond to incentives and punishments.[43] For these theorists, what is required to defeat an insurgency is a cost-benefit approach that disrupts and degrades the ability of the insurgent "system" to absorb resource "inputs" (e.g., recruits, weapons), process them, and produce "outputs" (e.g., political violence). By presenting the populace with a mix of carrots and sticks, the government seeks to structure the environment such that the rational choice for the populace is to act in ways that support the government.

Another school seeks to build a popular base that shrinks the political space in which the insurgents operate. In the struggle for the population, the government must demonstrate the capacity and the will to lead and must earn the public's allegiance by "winning hearts and minds," in the famous formulation of British counterinsurgency campaigners in Malaya. Developing physical infrastructure, operating under the rule of law, and promoting free and fair elections have been important elements in hearts-and-minds campaigns. Indeed, it is difficult to imagine a liberal democracy engaged in counterinsurgency, either directly or in a supporting role in another country, pursuing

---

[41] Taber (1974, p. 23).

[42] Laqueur (1976, p. 401).

[43] For the classic expression of this view, see Nathan Leites and Charles Wolf, Jr., *Rebellion and Authority: An Analytic Essay on Insurgent Conflicts*, Chicago: Markham, 1970.

a strategy that casts aside the hearts-and-minds approach in favor of purely repressive measures.

Both schools recognize that sterner measures must sometimes be employed if an insurgency is to be thwarted. These range from the surveillance and monitoring of the population (e.g., through the use of national identification cards, roadblocks, and random searches), through the creation of local militias, to the relocation of elements of the population to places easier to secure from insurgent penetration. Such efforts are risky, however, and can be self-defeating. For example, in Vietnam during the late 1950s, the Diem government uprooted thousands of peasants from their ancestral lands and forced them to build "agrovilles" at their own expense, frequently in inhospitable and remote parts of the country. This proved a disaster and greatly undermined peasant support for the government. If population control is to succeed—that is, if the insurgents are to be denied access to the food, information, weapons, and money they require—every care must be taken to ensure that the control regime is administered fairly, respectfully, and with restraint. Public information campaigns are essential to explain the government's goals, and grievance procedures must be in place to redress injustices honestly and expeditiously.

What about local militias and "counterorganizations"? Although widely recognized by analysts as a powerful counterinsurgency tool, local self-defense forces have sometimes been neglected or underutilized in practice. In an insurgency's early phases, conventional military forces often resist the creation of such organizations, arguing that they drain manpower and other resources from combat units.[44] Known by a variety of names—auxiliary police forces, village militias, home guards—these self-defense units, when properly organized, trained, and equipped, are useful not so much in terms of their physical power, which tends to be minimal, but in their ability to gather useful intelligence. As members of a given community, they have a far more intimate knowledge of local conditions than any government officials in the capital could hope to achieve and are likely to have a better understanding of local insurgent activities. Equally important, membership in a self-

---

[44] [British] Army (2001, p. B-7-A-1).

defense unit promotes political ties with the state and reinforces within its ranks a sense of common endeavor against the insurgents. Just as insurgencies, particularly under the classic Maoist variant, attempt to create parallel governments among the population, the government must build "counterorganizations" among the population. Organizing as much of the population into "parallel hierarchies" as possible was a technique the French employed usefully in Algeria during the 1950s. Among other things, these structures, which included not only militias but sports clubs, farmers' organizations, and veterans groups, gave colonial administrators a way to maintain close contact with disparate (and potentially restless) groups within Algerian society.[45]

Local self-defense units are ideally suited for a role in identifying and neutralizing an insurgency's underground command-and-control and support organization—the so-called "infrastructure" operating among the population but unseen by the state. The Viet Cong infrastructure, which engaged in recruitment, terrorism, subversion, and other clandestine activities, was the subject of a ferocious and effective campaign by U.S. and Vietnamese security forces during the late 1960s and early 1970s.[46] Today in Iraq, coalition forces are combating a "hydra-headed decentralized insurgency,"[47] consisting of an estimated 40 distinct groups.[48] Nevertheless, it seems likely that there is at least a rudimentary infrastructure in place that helps recruit, equip, and train fighters; gathers intelligence; and marshals funds.[49] Developing a fuller understanding of this invisible ecosystem is obviously the first step in

---

[45] John J. McCuen, *The Art of Counter-Revolutionary Warfare*, London, UK: Faber and Faber, 1966, pp. 98–100.

[46] Richard A. Hunt, *Pacification: The American Struggle for Vietnam's Hearts and Minds*, Boulder, Colo.: Westview Press, Inc., 1995, pp. 109–120.

[47] Ahmed S. Hashim, "The Sunni Insurgency in Iraq," *MEI Perspective*, Middle East Institute, August 15, 2003.

[48] Scott Johnson and Melinda Liu, "The Enemy Spies," *Newsweek*, June 27, 2005.

[49] According to press accounts, the intelligence on the insurgency has major gaps, which have "largely eluded the understanding of American intelligence officers since the fall of Saddam Hussein's government 27 months ago" (Dexter Filkins and David S. Cloud, "Defying U.S. Efforts, Guerrillas in Iraq Refocus and Strengthen," *New York Times*, July 24, 2005, p. 1).

disrupting it, and here local militias, with an intimate knowledge of the operating environment, are well placed to supply relevant information to police and intelligence services. Rather than separating the population from the insurgents, the emphasis here is on separating the insurgents from the population by removing the anonymity that allows them to operate freely.

4. Keep the Use of Force to a Minimum

This emphasis on targeting and neutralizing the insurgent infrastructure follows one of the cardinal tenets of modern counterinsurgency doctrine: the importance of keeping the government's use of force to an absolute minimum. Seeing the infrastructure as the critical target helps the government avoid the trap of relying on "search-and-destroy" operations designed to "find, fix, and fight" the insurgents, the default position of most states facing an armed rebellion. Insurgents rarely make themselves available for open combat with government troops and are often able to elude capture or destruction. Search-and-destroy operations are in fact worse than ineffective. The use of conventional combat power in populated areas easily leads to civilian casualties, the destruction of crops, and property damage and, in so doing, antagonizes the very population the government is seeking to enlist in its cause. Reliance on combat power overestimates its utility and "miscalculates the relevance of noncombatants and their attitudes to the outcome of the struggle."[50] As a leader of Algerian Front de Libération Nationale later explained, the French army's sweeps through populated areas, known as *ratissages* (literally, "rakings") were the insurgents' "best recruiting agent."[51]

It almost goes without saying that the police, paramilitary, and military forces engaged in maintaining the government's "control

---

[50] Edward E. Rice, *Wars of the Third Kind: Conflict in Underdeveloped Countries*, Berkeley, Calif.: University of California Press, 1988, p. 115.

[51] Quoted in Alistair Horne, *A Savage War of Peace: Algeria 1954–1962*, London, UK: Pan Books, 1997, p. 110.

regime"[52] must refrain from acting like an occupying force.[53] Strict standards of behavior must govern all interactions of the security forces with the civilian population. Again, a population antagonized by the depredations of the security forces is likely to provide many fresh recruits for the insurgent movement. At the very least, it is highly unlikely that that population will be willing to provide these forces with the intelligence essential to defeating the insurgency.

In this context, it is worth remembering that, while conventional wars certainly have political dimensions, counterinsurgency is far more politicized, and its practitioners must always be alert to the potential strategic consequences of even the most tactical of operations. It is the natural inclination of soldiers to display strength, courage, and aggression. But in counterinsurgency, "normal military logic is negated."[54] Otherwise admirable martial qualities can backfire in counterinsurgency's ambiguous environment, as Frank Kitson, a British army officer who led counterinsurgency campaigns in Kenya, Malaya, Oman, and Northern Ireland, has noted:

> Firm reaction in the face of provocation may be twisted by clever propaganda in such a way that soldiers find the civilian population regarding their strength as brutality, and their direct and honest efforts at helping to restore order as the ridiculous blunderings of a herd of elephants.[55]

Of course, this is not to suggest that combat power has no role in counterinsurgency. However, it has been most effective when conducted by forces highly trained in counterinsurgency techniques. For example, in rural insurgencies, combat power has proven most useful and appropriate when it is employed by small, specialized, highly trained counterguerrilla units capable of constant patrolling at long range and

---

[52] This phrase is taken from Gordon McCormick, "A Systems Model of Insurgency," unpublished paper, Monterey, Calif.: Naval Postgraduate School, June 22, 2005, p. 8.

[53] Blaufarb (1977, p. 310).

[54] Charles Townsend, *Britain's Civil Wars: Counterinsurgency in the Twentieth Century*, London, UK: Faber and Faber, 1986, p. 31.

[55] Kitson (1971, p. 200).

for extended periods, of ambushing and harassing the enemy, and of gathering intelligence that are backed up, when necessary, by "highly mobile army, air, and naval striking forces."[56] The government should also organize indigenous counterguerrilla units to collect intelligence and maintain pressure on the insurgents. Used with great effect against the Huk rebels in the Philippines, the Malayan Communist Party in Malaya, and against the Mau Mau in Kenya (1952–1960), "pseudo gangs" are essentially nonuniformed counterinsurgent forces disguised as guerrillas who operate (sometimes ruthlessly) within insurgent-dominated areas.[57] These techniques have worked well in rural societies but appear to have limited applicability in more-urbanized insurgencies, such as Iraq. Similar specialized forces and concepts do not appear to exist for urban counterinsurgency; they will need to be developed.

Increased urbanization adds another layer of complexity for military forces involved in counterinsurgency operations. Cities are extremely difficult operating environments for armed forces engaged in conventional combat.[58] Among other things, the complex physical, social, and political terrain of urban spaces often blunts the technological edge that helps characterize sophisticated military forces. Military units engaged in counterinsurgency operations in cities are likely to face even more daunting challenges. Minimizing civilian casualties, while an important goal of conventional U.S. operations in urban areas, is even more imperative in counterinsurgency, in which collateral damage can be a powerful recruitment and propaganda tool for the insurgents. Conventional military power will sometimes be needed to deal with an extremely violent insurgency, e.g., today in Iraq. But this will rarely be the case during an insurgency's formative period, when levels of violence are typically lower. During this earlier stage, police

---

[56] McCuen (1966, p. 121) and O'Neill (2005, p. 162).

[57] Lawrence E. Cline, *Pseudo Operations and Counterinsurgency: Lessons from Other Countries*, Carlisle, Pa.: U.S. Army War College, Strategic Studies Institute, June 2005, pp. 1–8, and John S. Pustay, *Counterinsurgency Warfare*, New York: The Free Press, 1965, p. 114.

[58] Alan Vick, John Stillion, Dave Frelinger, Joel Kvitky, Benjamin Lambeth, Jefferson Marquis, and Matthew Waxman, *Aerospace Operations in Urban Environments: Exploring New Concepts*, Santa Monica, Calif.: RAND Corporation, MR-1187-AF, 2000, p. xiv.

and paramilitary units are likely to prove more useful. While obviously capable of using deadly force when necessary, these units are typically trained to employ a much lower level of violence.[59] In addition, police (and to a lesser extent, paramilitary units), if well trained, are likely to have a closer relationship with local populations and are therefore more likely than combat units to develop human information sources. Indeed, in many respects, effective urban counterinsurgency, with its requirements for human intelligence, the minimum use of force, and developing relationships among the civilian population, is much closer to police work than it is to military operations.[60]

Border Control. Finally, combat forces, working with police and other civilian and paramilitary forces, can play a key role in preventing cross-border infiltration and arms shipments in support of the insurgency. In the early stages of an insurgency, weapons, food, and other resources can often be obtained locally, but as the movement grows, its requirements typically expand to the point at which outside assistance is necessary, either from state sponsors (e.g., in Kashmir today, where Pakistan supports numerous militant groups fighting Indian occupation forces) or from nonstate actors, such as diasporas (e.g., the Tamil communities in Canada that have helped finance the operations of the Liberation Tigers of Tamil Eelam [LTTE], the so-called "Tamil Tigers").[61] Moreover, while safe havens for rest, recuperation, and training can sometimes be found internally, insurgents must frequently rely on adjacent countries for sanctuary. To take one example, during the 1980s, Angola permitted Umkhonto we Sizwe [Spear of the Nation], the military wing of the African National Congress, to operate seven training camps inside the country, secure from attacks by the South African Defense Force.[62] Reliance on external support,

---

[59] For more on this point, see Grant Wardlaw, *Political Terrorism: Theory, Tactics, and Counter-Measures*, Cambridge, UK: Cambridge University Press, 1982, p. 98.

[60] Mounir Elkhamri, Lester W. Grau, Laurie King-Irani, Amanda S. Mitchell, and Lenny Tasa-Bennet, "Urban Population Control in Counterinsurgency," unpublished paper, Ft. Leavenworth, Kan.: U.S. Army Foreign Military Studies Office, October 2004, p. 1.

[61] See Chapters Two and Three of Byman, Chalk, et al. (2001).

[62] Photius Coutsoukis, "Angola as a Refuge," Web page, Rev. ed., November 10, 2004.

whether in terms of materiel, international propaganda, intelligence, or sanctuary, imposes costs on the insurgents. It can limit their freedom of action, since states or diasporas can impose unwelcome conditions in exchange for assistance. In addition, external support can lead to a dangerous dependency on the continued good will of outsiders. That good will can evaporate quickly, and with lethal consequences, as insurgents in Iraqi Kurdistan discovered in 1975, when the Shah of Iran suddenly halted all shipments of supplies to the movement and shut down his country's safe havens.[63]

The ongoing insurgency in Iraq has introduced a new variation on the problem of border control. As they battle coalition and Iraqi forces, indigenous and foreign fighters are gaining invaluable experience in guerrilla warfare, terrorism, and urban combat. U.S. officials worry that the war is forging a new generation of highly professional insurgents and terrorists and that these fighters will "bleed out" of the country and join other conflicts, much as the mujahideen who fought Soviet forces in Afghanistan during the 1980s went on to fight in Bosnia, Chechnya, and Kashmir in the 1990s.[64] Iraq demonstrates that border control must aim at more than keeping foreign fighters, weapons, and other resources out but must also be part of a broader strategy aimed at preventing the influx of newly professionalized jihadists into other conflicts.

Need for Rectitude. Although the insurgency may have reached the point of crisis, at which the embattled government is fighting for its very existence, it is imperative that the security forces, indeed for all representatives of the state, to "display rectitude" in their dealings with the insurgents.[65] Treating prisoners fairly, offering amnesty when appropriate, and resettling former fighters who have "rallied" to the government's side can create incentives for members to leave the militant movement, thus helping to deplete that movement's ranks. In so

---

[63] Gerard Chaliand, *Terrorism: From Popular Struggle to Media Spectacle*, London, UK: Saqi Books, 1987, p. 58.

[64] National Intelligence Council (2004); Douglas Jehl, "Iraq May Be Prime Place for Training of Militants, CIA Report Concludes," *New York Times*, June 22, 2005, p. 10.

[65] Joes (2004, p. 237).

doing, these measures contribute to a broader political objective, as demonstrated with the Chieu Hoi [Open Arms] program during the Vietnam War. According to one assessment of the program, Chieu Hoi

> [n]ot only weakened the enemy's manpower by the defection of over 194,000 ralliers . . . [but also] weakened his cause politically with the public rejection of the policies and objectives of the National Liberation Front by a substantial number of its former adherents.[66]

Conclusion

As in other political-military struggles, counterinsurgency requires persistence, resilience, and the ability to adapt rapidly to changing circumstances and conditions. But the differences between counterinsurgency and conventional war are probably more profound than are the similarities. For example, the adversaries in this unconventional conflict have (or should have) different standards for success. For the insurgents, long-term survival is itself a form of victory in that it demonstrates the continuing impotence of the regime or occupying power.[67] For the counterinsurgent, outright "victory" is rarely achievable, Malaya and a handful of other campaigns serving as the exceptions that prove the rule. Typically, success is nothing more grandiose than containing the insurgent movement and reducing the conflict to what British Home Secretary Reginald Maulding, in his famous comment referring to Northern Ireland, termed "an acceptable level of violence."[68]

---

[66] J. A. Koch, *The Chieu Hoi Program in South Vietnam, 1963–1971*, Santa Monica, Calif.: RAND Corporation, R-1172-ARPA, 1973, p. v. Recently in Iraq, political leaders have discussed amnesty for insurgents but have ruled out any form of clemency for foreign fighters operating there. Ellen Knickmeyer, "Iraq Proposes Broader Amnesty," *Washington Post*, April 11, 2005, p. 1.

[67] Taber (1974, p. 30). Donald Snow makes a related point when he argues "insurgencies win by not losing" (Snow, 1996, p. 75).

[68] Quoted in Bruce Anderson, "The Enemy Within," *Spectator* (London), October 6, 2001, p. 30.

No two insurgencies are identical, and thus there can never be a universally applicable template for delivering success. As one leading scholar, Walter Laqueur, has observed, "[r]eality is always richer and more complicated than any formula, however ingenious."[69] That said, some basic requirements for counterinsurgency have remained constant.[70] These include extending state capacity and presence; controlling the population; minimizing the use of force; and, perhaps most important, developing as truthful an understanding of the adversary as possible.

Prevailing over an insurgency also requires a threatened government to conduct honest assessments of its own political, economic, military, and social strengths and shortcomings. In a regime with a highly politicized, corrupt, or incompetent officer corps or civil service, candid self-appraisals will likely be rare. Nevertheless, outside powers supporting the embattled government should encourage rigorous self-analysis as part of a broader "self-help" approach to combating the insurgency. Similarly, outside supporters should also conduct their own rigorous appraisals of their client's capabilities and be unflinchingly honest in dealing with the implications of such reviews, no matter how badly they may reflect on the host nation.

As useful as these principles are, they are not sufficient to deliver success in counterinsurgency. Despite the fact that the U.S. government has conducted counterinsurgency operations since Reconstruction, when the U.S. Army fought southern counterrevolutionaries opposed to rights for African-Americans, counterinsurgency has been of fleeting interest to policymakers.[71] One consequence has been the failure to build institutional capacity within the U.S. government for planning, coordinating, and carrying out counterinsurgency activities by civilian, as well as military, organizations. For example, while service and joint doctrine exists for counterinsurgency, there is no interagency doctrine, despite near-unanimous agreement that successful

---

[69] Laqueur (1976, pp. 378–379).

[70] Beckett (2005, p. 15).

[71] Andrew J. Birtle, *U.S. Army Counterinsurgency and Contingency Operations Doctrine, 1860–1941,* Washington, D.C.: U.S. Army Center for Military History, 2001, p. 57.

counterinsurgency requires a range of nonmilitary tools. As evidenced in the early days of the insurgency in Iraq, when the U.S. military, the Coalition Provisional Authority, and the intelligence community sometimes operated at cross-purposes, coordination among agencies remains a major challenge. Neither is any formalized "early warning" system in place for identifying nascent insurgencies.

Such shortfalls will have to be addressed, particularly if the United States expects to be waging a protracted counterinsurgency campaign on a global scale. In the meantime, it is useful to remind ourselves that successful counterinsurgency is unlikely to come easily. Most insurgents fail, but some have been remarkably effective against the world's most capable powers. During the second half of the 20th century, as Martin van Creveld reminds us,

> [f]rom France to the United States, there has scarcely been one "advanced" government in Europe and North America whose armed forces have not suffered defeat at the hands of under-equipped, ill-trained, ill-organized, often even ill-clad, underfed, and illiterate freedom fighters or guerrillas or terrorists; briefly, by men—and, often, women, who were short on everything except high courage and the determination to endure and persist in the face of police operations, counterinsurgency operations . . . and whatever other types of operations were dreamt up by their masters.[72]

In the next chapter, we take a step back and consider counterinsurgency from the standpoint of grand strategy, discussing how great an effort the United States should devote to the problem, as well as the relative advantages of alternative strategies.

---

[72] Martin van Creveld, *The Rise and Decline of the State*, Cambridge, UK: Cambridge University Press, 1999, p. 395.

CHAPTER FOUR

Grand Strategy and Counterinsurgency

Introduction

As a result of the insurgencies that the United States and its allies have been combating in postinvasion Afghanistan and Iraq, counterinsurgency has once again become a central concern in U.S. defense planning. However, while it is generally accepted that counterinsurgency operations and capabilities will be important in the future, the amount of emphasis that they should receive relative to other priorities for military attention and investment remains an open question—arguably the most important one for U.S. policymakers to answer in designing military forces for the next several decades.

This chapter considers the questions of how great a role preparing for counterinsurgency should play in U.S. strategy making and of what shapes such preparations might take, issues into which the subsequent chapters of this study then delve further. These are matters of grand strategy, on two levels.[1] First, the extent to which the United States should focus its defense energies on counterinsurgency is an intrinsically strategic question, since the answer depends on assessing the full range of security threats the United States faces and setting priorities among the many possible policy responses to them. Second, counterinsurgency is a particularly grand strategic problem in its own right, as the preceding chapters have described, because it involves both

---

[1] *Grand strategy*, in Barry Posen's apt and enduring definition, is "a political-military means-ends chain, a state's theory about how it can best 'cause' security for itself." Barry R. Posen, *The Sources of Military Doctrine*, Ithaca, N.Y.: Cornell University Press, 1984, p. 13.

military and nonmilitary policy instruments, both of which are usually essential to success, and the effective integration of which has been a critical, enduring challenge in past conflicts.

Assessing Insurgent Threats to U.S. National Security

How much attention the United States should devote to preparing for and carrying out counterinsurgency campaigns depends on answering two basic questions. The first, more obvious of these is how great a threat insurgencies should be expected to pose to U.S. national security interests in the future. This involves comparing this security threat to others but also assessing the threat from insurgency in absolute terms: Even the worst threat a secure state faces may be mild, while in perilous times a threat may be dire yet still be less urgent than others that appear even more dangerous. The second factor, more easily overlooked, is how much the United States can do to avert or reduce these threats: Even if a problem is severe, it makes little sense to invest heavily in ineffective or exorbitantly costly responses to it. The following discussion provides a framework for making such assessments but, given the nature of this study, does not purport to offer definitive answers to these questions.

When Do Insurgencies Threaten U.S. Security Interests?

As Chapter Two of this study describes, there are dozens of insurgent groups and movements in the world, even setting aside the uncountable host of aspiring or would-be insurgencies that are too small or weak to be more than a nuisance to the states they seek to afflict. Many of these pose little or no threat to U.S. interests. For example, although the LTTE in Sri Lanka is arguably the most capable insurgent group in the world, it is not aggressively opposed to the United States. Substantial intervention against LTTE has thus never been a tempting policy option, even though Washington would certainly welcome the end of the group's war against the Sri Lankan government.[2]

---

[2] However, the Tamil Tigers have developed terrorist tactics and techniques (most notably in the field of suicide attacks) that other groups have later adopted and used against the United States and its allies.

During the Cold War, U.S. assessments of the dangers posed—or the opportunities presented—by particular insurgencies were generally a function of whether the insurgents or the government they were attacking was friendly to Moscow or to Washington and how important the territory over which they were fighting was to the superpowers. In today's less-polarized world, there are four general conditions under which an insurgency poses a substantial security threat against the United States, although U.S. intervention will not necessarily appear to be worthwhile in every such case.

First are conflicts in which the insurgents are dangerously hostile to the United States or in which it appears that the insurgents might win and that the regime that would result from an insurgent victory would be dangerously hostile. Although, in theory, many sorts of insurgents might be both powerful and actively anti-American, in practice, most of the insurgencies that currently pose this sort of a direct threat to U.S. security involve Muslim extremist groups that would conduct, spawn, or facilitate terrorist activity directed against the United States. Among the exceptions to this generalization are nationalist insurgents attacking U.S. forces and U.S.-supported governments in Iraq and Afghanistan (although separating the nationalist from the Islamist insurgents in these states is problematic) and perhaps FARC in Colombia, which is not particularly belligerent toward the United States today but might become so if it took control of the country.

A second category of dangerous insurgencies comprises cases in which the insurgents prevent the government from exercising control of territory that anti-U.S. terrorists can then use as a base of operations, for example in such nations as Georgia and Somalia. It is important to note that insurgencies often prevent states from being able to police parts of their territory, both by contesting control of that territory and by draining government resources, but by no means do terrorists always exploit this loss of control. Whether intervening to prevent such power vacuums is worthwhile will depend on how potentially useful the area would be to terrorists and on the expected costs of trying to avert the problem.

The third threatening class of insurgencies consists of those that imperil or seriously weaken important U.S. allies. For example, the

Irish Republican Army was not hostile to the United States—indeed, individual Americans provided it with considerable support—but its attacks against the British government drained strength from the Western alliance by weakening one of its principal members. Israel has faced similar threats from Palestinian insurgents. It is rare for important U.S. allies to be fragile enough for insurgents actually to threaten their survival; some states that are strategically critical to current counterterrorist efforts could become vulnerable to direct insurgent overthrow, but these are of concern primarily because of their potential to become hotbeds of terrorism, as discussed above.

Finally, and related to the preceding category, are insurgencies whose success, or even mere survival, threatens to destabilize other states that matter to the United States, either by fomenting insurgency in neighboring states or simply by serving as an inspiration for it. Fears that insurgency would be contagious were common, though often overblown, during the Cold War. Today the specter of dominoes falling to Marxist national liberation movements has been replaced by concerns that the success of Islamist groups in one state will inspire and encourage their counterparts in others. Although exporting revolution across national borders is generally more difficult in practice than it appears to be on paper, there is little question that such inspirational effects do occur, as in the case of the U.S. withdrawal from Somalia—like the earlier Soviet defeat in Afghanistan—encouraging insurgents facing American forces in other places to think that a strategy of attrition against U.S. forces will eventually bring success.

Although this is a long list, many insurgencies fall outside these categories, as in the case of Sri Lanka and the LTTE. Moreover, counterinsurgency intervention is not necessarily called for in every case that does fit one of these criteria: Sometimes, U.S. intervention will do little to enhance the prospects of defeating or containing the insurgents or will even threaten to make the situation worse. This may be true because the local government is fully capable of handling the problem on its own or because there is little prospect that U.S. involvement will improve a situation that is essentially hopeless, at least at a reasonable price.

On the other hand, the United States may also choose to intervene against insurgencies that do not threaten U.S. security interests but where there are powerful humanitarian or other reasons to do so, for example, to prevent insurgencies leading to mass killings like those in Cambodia and Rwanda. While interventions in such cases are likely to be infrequent, at least when powerful insurgent movements are involved, the less common that truly dangerous insurgencies are, the easier it will be to contemplate such optional interventions, for the simple reason that the forces and resources needed to conduct them will be subject to fewer competing demands.

The Importance of Insurgency and Counterinsurgency in Future U.S. Grand Strategy

Having considered the circumstances under which particular insurgencies pose threats to U.S. national security, how significant a threat should we anticipate that insurgencies in the aggregate will pose in the future? For military planning, this question is often posed in terms of the relative importance of conventional and counterinsurgency operations in future conflict: To the extent that there are trade-offs to be made between investing in conventional and counterinsurgency capabilities, what is their relative importance?

It is relatively easy to declare that insurgent threats have grown in importance relative to conventional ones in recent years and that this situation is not likely to change in the near future, for the simple reason that the number of serious conventional military threats to U.S. interests has declined substantially. With Iraq removed from this category since 2003, at most only Iran, North Korea, and China (against Taiwan) appear to pose potentially major conventional threats, and dealing with the last of these is a problem almost entirely for naval and air power. Not only is this a far more limited constellation of threats than characterized the Cold War world, it is also more limited than the set of conventional threats for which the United States had to prepare a decade ago, when Iraq and Serbia were also numbered among the dangerous international rogues.[3]

In contrast, current and potential insurgent threats remain important, not only in postinvasion Iraq and Afghanistan but also in a variety of other locations, a pattern that is due primarily to two interconnected developments: the emergence of modern transnational terrorism and the rise of extremist Islamism. As the potential destructive capability of terrorists increases, bases from which they might operate become more serious threats to national security, and insurgencies may provide such bases, either because the insurgents are sympathetic to the terrorists' cause (or, like the Taliban, beholden to the terrorists) or because they create territorial power vacuums in which governments cannot effectively prevent terrorist operations. In addition to their relationship to terrorism, radical Islamist insurgent movements hold the potential to destabilize a number of states that the United States counts as important allies, or at least cobelligerents, in its counterterrorist efforts and to spill over borders even from countries whose fate is not intrinsically important to Washington.

In this sense, the aggregate insurgent threat resembles the one that U.S. leaders perceived from procommunist insurgencies during the Cold War, but there are important differences between the two cases. Most obviously, the constellation of insurgencies of concern to the United States today is less centralized and monolithic than the one insurgents presented during the Cold War—and vastly less so than the insurgent threat that Washington perceived at the time.[4] Many contemporary insurgencies receive important external support, but instead

---

[3] Of course, recent or impending development of nuclear weapons by Iran and North Korea makes these states more potentially dangerous than they once were. It does not, however, increase their conventional military capabilities or necessarily make them more difficult to contain.

[4] During the Cold War, U.S. leaders greatly overestimated the threat from third-world insurgencies, which individually proved to be less useful and more expensive to Moscow than was widely feared and which collectively constituted something far short of an ideologically unified alliance committed to the defeat of the West. It is still possible that much the same might prove to be true of the Islamist extremists, whose movement could yet splinter into competing factions and ultimately collapse under its own inefficient weight. Prudent planners will not count on this happening, however.

of a distant superpower, it is neighboring states, nonstate actors, or ethnic diasporas that tend to provide this aid.[5] The insurgencies themselves vary more widely as well, with only a few conforming to traditional Maoist models (see Chapter Two). This diversity of insurgencies makes it even more true now than in earlier eras that no "one-size-fits-all" approach to counterinsurgency will be adequate, compelling U.S. strategists to examine each case on its own terms in order to develop policy responses that will provide appropriate assistance to threatened governments while minimizing the potential for U.S. involvement to backfire.

Grand Strategies for Small Wars

The question of how important counterinsurgency will be among future U.S. military operations would not matter for force planning purposes if preparing for conventional and counterinsurgency operations called for essentially the same investments in force structure, training, and education. However, counterinsurgency is not simply a "lesser included case" of conventional warfare, as both the U.S. Army and Air Force once assumed it to be.[6] Instead, conducting counterinsurgency well calls for a set of capabilities that is different in many important respects from those that are optimal for conventional combat against traditional opponents.

Differences Between Counterinsurgency and Conventional Military Requirements

The most fundamental difference between counterinsurgency and conventional warfare, and the one from which most of the others derive, is that the contest between insurgents and counterinsurgents is primar-

---

[5] Byman et al. (2001).

[6] Army doctrine still declares this to be the case: "The doctrine holds [conventional] war-fighting as the Army's principal focus and recognizes that the ability of Army forces to dominate land warfare also provides the ability to dominate any situation in military operations other than war" (U.S. Army, *Operations*, Washington, D.C.: Headquarters, Department of the Army, FM 3-0, June 14, 2001, p. vii).

ily conducted on a political rather than a physical battlefield, where the outcome is determined by whether the insurgents or the government ultimately wins the predominant support of the populace. From this flows the need for direct, sustained personal contact between the troops conducting counterinsurgency and the local populace, not only because of the need to collect human intelligence, conduct searches, and perform other tasks that cannot be done remotely but also to provide security, create confidence in the government, and build the relationships that win hearts and minds at the grass-roots level.

The enduring labor intensiveness of counterinsurgency, with its requirement for many literal "boots on the ground," stands in contrast to the increasing substitution of capital for frontline labor in the evolution of conventional military forces; technological advances in mobility, stand-off sensors, command-and-control networks, and precision attack make it possible for small numbers of aircraft, vehicles, or troops to perform combat missions that would have required much larger forces a generation ago.[7] This is not to suggest that technological progress in these areas does not enhance the ability of modern armed forces to conduct many of the tasks required in counterinsurgency operations. However, the optimal force mixes for warfare against regular and irregular adversaries differ even more today than they did in the 1960s, when the forces designed for fighting the mechanized armies of the Warsaw Pact on the European central front proved to be ill-suited for conducting counterinsurgency against the Viet Cong, and vice versa. This is well illustrated by the contrast between the initial phases of Operation Iraqi Freedom (OIF), when aerial firepower enabled relatively small, highly mobile ground forces to advance to Baghdad with little difficulty, and the requirement for the sustained presence of large numbers of relatively light ground forces to conduct the subsequent counterinsurgency campaign.[8]

---

[7] Bruce R. Pirnie, Alan J. Vick, Adam Grissom, Karl P. Mueller, and David T. Orletsky, *Beyond Close Air Support: Forging a New Air-Ground Partnership*, Santa Monica, Calif.: RAND Corporation, MG-301-AF, 2005, Chapter Two.

[8] See David E. Johnson, *Learning Large Lessons: The Evolving Roles of Ground Power and Air Power in the Post–Cold War Era*, Santa Monica, Calif.: RAND Corporation, MG-405-AF, 2006.

Because of the need for contact with the local population—along with the tactical considerations associated with fighting an adversary that operates primarily in dispersed units in urban and complex terrain and intermingled with the civilians whose support and confidence the counterinsurgency effort seeks to gain and maintain—counterinsurgency is by its nature primarily the domain of ground forces. Air power, for example, makes a greater proportion of its contribution to counterinsurgency by providing intelligence, surveillance, and reconnaissance and airlift than it does in conventional warfare.[9] However, it is easy to overstate the extent to which this is true, especially when looking back at the effects of air attack against insurgents in canonical historical cases, such as Malaya and Vietnam, in light of the limited but important improvements that have occurred in the ability of aircraft to strike such targets. Today, aerial firepower has become the clear instrument of choice for providing fire support for ground forces operating in areas where collateral damage must be minimized.

The effects that joint, ground-centric combat forces seek in counterinsurgency are less physical and more social than is usually the case in conventional warfare. Providing security for the populace through policing and, when required, larger-scale counterinsurgency combat operations is central to successful counterinsurgency efforts, but maximizing effectiveness on the political battlefield of insurgency also calls for heavy emphasis on developing and employing both military and nonmilitary capabilities for psychological operations, public information, civil affairs, and other "nonkinetic" functions. Similarly, it is often valuable to use military capabilities in such fields as medicine and civil engineering to provide benefits to civilians, not so much to bribe them into supporting the government as to help persuade them that it has their welfare at heart, which increases the desirable amount of such capabilities in the counterinsurgency force well beyond what is required merely to support the combat forces.

Identifying the sorts of capabilities that are useful in a counterinsurgency campaign is not the same thing as specifying what capa-

---

[9] James S. Corum and Wray R. Johnson, *Air Power in Small Wars: Fighting Insurgents and Terrorists*, Lawrence, Kan.: University Press of Kansas, 2003.

bilities the United States should provide in counterinsurgency interventions, however. As we note elsewhere in this monograph, only the actions of the local government can demonstrate its good intentions; civic action and other activities by outsiders may be beneficial but do not accomplish that essential goal. What the United States must provide will, of course, vary from one case to another, depending on what the local government cannot provide for itself, in what areas its capabilities can be substantially improved by external assistance, and what the political consequences of any particular foreign assistance or presence are likely to be. In each case, the ultimate objective is shifting the balance of power—including both military power and political legitimacy—between the insurgents and the government decisively in favor of the latter.

Differences Between Counterinsurgency and Conventional Military Strategy

The same factors that shape the military capabilities required for counterinsurgency also have important effects on the nature, or at least the substance, of strategy-making for counterinsurgency. All military strategy is, or should be, fundamentally political, as Clausewitz teaches even those who know little else of his work.[10] However, conventional warfare still tends to be an arena in which the military contest, narrowly defined, is the central matter of concern, around which economic, diplomatic, and other policies revolve. In short, military strategy becomes the centerpiece of grand strategy. For insurgents, and thus for those who fight against them, military strategy is subordinate to political strategy in a far more practical, immediate way. Counterinsurgency strategy is grand strategy in miniature, and to consider its military component in isolation is as artificial as examining the air component of a conventional joint military campaign without regard to the effects of surface forces.

This has many practical implications, such as the need for military officers preparing for counterinsurgency to become conver-

[10] See Carl von Clausewitz, *On War*, trans. Michael Howard and Peter Paret, eds., Princeton, N.J.: Princeton University Press, 1976 [1973].

sant with the other instruments of power and adept at collaborating with the other agencies and organizations—both U.S. and foreign—alongside which they will operate.[11] This is important whether the United States is engaging in large-scale counterinsurgency warfare or providing much smaller increments of assistance to governments facing nascent insurgent threats with the goal of preventing the larger conflicts from developing.

One of the most distinctive—and challenging—features of making and executing counterinsurgency strategy is that "information operations" are central to strategic success, not merely an appendage of kinetic operations. It would be unfair to say that the U.S. armed forces pay mere lip service to the importance of propaganda, public affairs, psychological operations, and the other elements of information operations, but it is certainly true that these elements of military power are generally treated as useful appendages of the military apparatus, whose value is usually measured merely in terms of their ability to facilitate or enhance traditional combat functions.

The political battlefield that counterinsurgency operations must seek to dominate is one in which perceptions and beliefs are what matter. If people believe that U.S. forces conducting counterinsurgency are attacking civilians indiscriminately, that their government is irredeemably corrupt, or that their streets are unsafe, it is irrelevant to the conflict whether these things are actually true except insofar as it may affect these impressions. Particularly as information technology grows ever more sophisticated, virtually any event on the physical battlefield can have significant political consequences, and it is difficult to overstate the value of being able to shape perceptions of these events in ways that advance the overall strategy, and to prevent the enemy from doing the same.

Finally, counterinsurgency strategy must accommodate the fact that insurgent warfare involves prolonged conflict, requiring both

---

[11] The same imperative exists for relevant members of the other entities, of course, and tends to be even more challenging for them to achieve, for reasons ranging from organizational culture to the lack of analogues to the professional military education system for the State Department or the Central Intelligence Agency.

patience and adaptation. Popular doctrinal concepts, such as "rapid decisive operations" that emphasize the merits of winning quick victories or achieving strategic-level shock and paralytic effects against conventional military opponents are intrinsically inconsistent with this reality, and seeking to apply them in counterinsurgency will often be counterproductive. Defeating any substantial insurgency requires sustained effort over years or even decades, which affects everything from planning for appropriate force rotation to managing expectations when the conflict is presented to domestic political audiences at home.

Coercion in Counterinsurgency

As in conventional military conflict, successful counterinsurgency usually involves coercion: convincing the enemy that it is better to surrender—or in the case of the population, to switch allegiance to the government—than to continue fighting.[12] Insurgencies generally do not end because the last insurgent has been killed.[13] Rather, they end when the movement's leaders, members, or supporters decide that the conflict is no longer worthwhile, either because it is too costly (or, for the individual, too dangerous), because success appears hopeless, or both—or they end when what is left of the government decides to concede, and the insurgents declare victory.[14]

---

[12] See David E. Johnson, Karl P. Mueller, and William H. Taft V, *Conventional Coercion Across the Spectrum of Operations: The Utility of Military Force in the Emerging Security Environment*, Santa Monica, Calif.: RAND Corporation, MR-1494-A, 2002, Chapter Two; Daniel Byman and Matthew Waxman, *The Dynamics of Coercion: American Foreign Policy and the Limits of Military Might*, Cambridge, UK: Cambridge University Press, 1992; Karl P. Mueller, "The Essence of Coercive Air Power: A Primer for Military Strategists," *Royal Air Force Air Power Review*, Vol. 4, No. 3, Autumn 2001, pp. 45–56.

[13] Exceptions most often occur against very weak enemies that can be utterly destroyed relatively inexpensively, or (rarely) against truly extreme opponents who remain committed to fighting no matter how high the costs and how remote the prospect of success. On the relationship between coercion and brute force, see Thomas C. Schelling, *Arms and Influence*, New Haven, Conn.: Yale University Press, 1966, Chapter 1.

[14] As suggested here, the two basic approaches to coercion, which includes both deterrence and compellence, are *punishment* (seeking to make the costs of misbehavior prohibitively high) and *denial* (making misbehavior appear fruitless). See Robert A. Pape, *Bombing to Win: Air Power and Coercion in War*, Ithaca, N.Y.: Cornell University Press, 1996, Chapters 2–3;

The problem of coercing violent nonstate actors, such as insurgent groups, is easy to write off as hopelessly difficult. Many of the punitive coercive levers commonly employed against states find little purchase against enemies with few tangible assets to threaten and may indeed drive the populace more firmly onto the side of the insurgents. The task can also be frustrating because victory is often less absolute than it is in wars between states: As Chapter Three explained, in counterinsurgency, as in counterterrorist operations, success often means not the end of insurgent violence but merely its reduction to a tolerable level. However, the reality is that, in general, coercing insurgents is not impossible but merely complicated and challenging—as noted in Chapter Two, most insurgent groups ultimately do give up. Of course, this general pattern does not mean that every insurgency can be defeated by its respective government if only the latter adopted an optimal strategy.

Several important differences between modern states and typical insurgent groups may present distinctive challenges to coercion strategists in counterinsurgency.[15] The first is that, when coercing groups that cannot take the loyalty and participation of their constituents more or less for granted, as most modern states can, coercion becomes a multilayered game in which the coercer typically seeks to alter behavior, both that of the target group as a whole and that of individual members, supporters, or those who might join or support the group.[16] Insurgents who no longer enjoy at least the tacit support of a segment of the population sufficient to sustain and, when necessary, help con-

---

Karl Mueller "Strategies of Coercion: Denial, Punishment, and the Future of Air Power," *Security Studies*, Vol. 7, No. 3, Spring 1998, pp. 182–228.

[15] The same considerations apply in many cases to efforts to coerce other violent nonstate actors, such as terrorist groups.

[16] To this list of potential coercion targets could be added, in some cases, foreign states or other supporters of the insurgency. Obviously, there is considerable variation in the extent to which the members and local populations of insurgent groups—and states and even armies, for that matter—have the freedom or inclination to make independent choices about whether to follow the group's leaders. Some insurgent groups are even more effective at mobilizing their followers than are many nation-states.

ceal them cannot survive even against a relatively weak counterinsurgent government.

Some policies may work on both the group and individual levels at the same time: For example, a denial strategy that makes victory for the insurgents appear impossible might simultaneously encourage the group's leaders to moderate their objectives; cause individual insurgents to abandon the fight; and discourage potentially supportive citizens from joining, helping, or protecting the insurgents. In other cases, however, one of the target audiences may be susceptible to incentives or threats that others are not. For example, if the insurgent leaders are motivated by an extremist ideology with essentially unlimited aims but the movement's rank and file seeks more-limited goals, coercive measures (including government concessions) may be able to weaken the insurgency even without altering the determination of the hard-core leadership.[17] Alternatively, it may be possible to buy off venal leaders, such as some Afghani warlords in past conflicts, with personal rewards, doing nothing to satisfy the grievances of their followers but stripping the insurgency of its direction and coordination or of significant allies.[18]

Many, but not all, violent political groups have organizational characteristics that can make them respond to coercion in ways not typically associated with solidly institutionalized states. In particular, the common assumption that states will behave in at least loosely rational ways is based on expectations that leaders will achieve their positions only if they possess some degree of intelligence and sanity; that their freedom to govern by whim will be at least minimally constrained by advisers and supporters; that, during crises and confrontations, they will be provided with reasonably accurate information about the situa-

---

[17] Similar conditions can exist if the leadership of an insurgency values being insurgents essentially for its own sake, but the movement depends on the support of followers who are interested in achieving concrete successes and will be discouraged if these are not achieved, as arguably became true for the Irish Republican Army in recent decades.

[18] Although often ignored in discussions of coercion, promises of reward, also known as *positive sanctions*, work in essentially the same way as threats of punishment and frequently offer significant advantages over them. See David A. Baldwin, "The Power of Positive Sanctions," *World Politics*, Vol. 24, No. 1, October 1971, pp. 19–38; and Leites and Wolf, 1970.

tion; and that, once national policy decisions are made, they will tend to be put into practice by those who are responsible for doing so. These assumptions are not always true in practice even for states, but major exceptions to them are relatively rare.[19] Among nonstate actors, the chance of behavior that departs dramatically from these expectations tends to increase to the extent that the groups are very small or highly decentralized, are led by autocratic individuals, or are seriously lacking in bureaucratically rational institutions. As described in Chapter Two, insurgencies vary widely in organization and ideology, so highly idiosyncratic behavior would be less surprising among, say, "lumpen insurgents" like the Revolutionary United Front in Sierra Leone or the cultish Lord's Resistance Army in Uganda than it would be on the part of the more institutionalized FARC or LTTE. Such tendencies will not necessarily make a group harder to coerce—irrationality can cut either way—but they may increase the difficulty of predicting coercive outcomes in the absence of detailed knowledge about the insurgent group in question.[20] Weakly organized insurgent groups may also complicate the coercion process if followers do not reliably obey their leaders' policies and instructions.[21]

An additional feature of many insurgencies that is important for would-be coercers to take into account is desperation. People who believe that they have little to lose—that is, that complying with the government's demands is so unattractive that it makes the costs of even fruitlessly continuing or supporting the fight appear worthwhile—are

---

[19] Saddam Hussein's remarkably dysfunctional decisionmaking prior to the 2003 invasion of Iraq, based on exceptionally poor information and a highly unrealistic understanding of his adversary, probably represents the greatest departure from most of these assumptions by the government of a significant state in recent decades.

[20] It is important to note that organizational rationality is a separate issue from the question of whether a group's goals or ideology is rational. Insurgent groups rarely exist to pursue objectives that are irrational in a formal sense, since it is hard to enlist people to join organizations whose purposes make no sense to them. Outsiders and opponents may consider their objectives or actions to be profoundly foolish or incomprehensibly misguided, and the organization or its members may in practice be motivated by goals other than those for which the movement ostensibly stands, but the need to understand how the world looks through the enemy's eyes is familiar to good coercion strategists.

[21] See Byman and Waxman (1992, pp. 190–193).

extremely difficult to coerce through either denial or punishment. Because participating in an insurgency is typically very dangerous, especially in its early stages, those who do so will frequently be unpromising targets for coercive threats. Thus, as in El Salvador and Nicaragua in the 1980s, successful coercion that makes settling the conflict appear attractive may depend on supplementing denial and punishment with positive sanctions, such as offering economic or political reforms, power-sharing arrangements, amnesties, or (in the case of secessionist movements, such as the Bosnian Serbs in the 1990s) limited local autonomy.

Conversely, insurgents who are motivated by the prospect of achieving or advancing an especially alluring "revolutionary dream" will be difficult to coerce because it may appear to make no price too high to pay and no risk too much of a long shot to be worth taking. Thus again, punishment and denial may both be weak approaches. Instead, successful coercion may hinge on reducing the appeal of the dream, either by eliminating the belief that insurgent victory would actually bring the promised rewards if it were achieved or by fostering the idea that there is an alternative path to attaining the desired benefits that is more promising than insurgency.[22]

Actually undermining such a motivating ideology is rarely easy and is often exceptionally difficult. This is particularly true when the insurgents' motivation is based on powerful religious beliefs: If people believe that their government is actually *evil*, not merely incompetent or oppressive, persuading them to shift their allegiance to it and away from the insurgents should require changing their most fundamental beliefs. Such changes do occur—witness the very gradual disappearance of religious warfare in Europe during the second millennium—but they are not easily brought about and rarely happen quickly. This is perhaps the greatest challenge that the United States faces today in

---

[22] Threatening to deprive the enemy of the fruits of victory, rather than make victory look impossible, has some features in common with threats of punishment but in this case would involve political rather than physical means. Offering an enemy an alternative way to achieve some of its goals usually involves positive sanctions, such as offers of inclusion in the political process. See Karl P. Mueller, *Strategy, Asymmetric Deterrence, and Accommodation*, Ph.D. diss., Princeton University, 1991, Chapter 2.

dealing with the threat of a transnational insurgent movement united by radical Islamist ideology.

Options for Counterinsurgency Intervention

Counterinsurgency and conventional military operations differ profoundly at the strategic, operational, and tactical levels, which military planners forget only at their peril. However, it is equally important to recognize that not all counterinsurgency operations are alike.

The Diversity of Counterinsurgency

When we speak of U.S. "intervention" against insurgents, it tends to evoke images of large-scale military operations against powerful, irregular enemies in the jungles of Vietnam, the mountains of Afghanistan, or the cities of Iraq. Yet these campaigns and others like them represent only the most visible—and the most expensive—end of a much larger spectrum of responses to insurgent threats.

Counterinsurgency intervention, as the term is used here, can take many forms. Politically, the United States may play a leading, secondary, or only minor role in relation to that of the local government and in some cases other outside powers. The intervention may be unilateral; formally or informally multinational; or conducted under the auspices of an intergovernmental organization, such as the United Nations. In some cases, U.S. participation in the conflict may be primarily military; in others, the armed forces may be only a small part of a U.S. policy package in which economic, intelligence, police, or other nonmilitary assistance has far greater influence.[23] U.S. armed forces may fight the insurgents themselves, support the combat operations of local forces directly or indirectly, or limit their role to training and advising government forces that conduct operations on their own.

---

[23] By extension, we could label as counterinsurgency intervention some policies to assist states facing insurgencies in which no U.S. military forces at all are involved, but this study limits its attention to cases in which armed forces play at least a small role in the intervention.

Ideal Types: Precautionary and Remedial Counterinsurgency

When facing this wide range of variation on many dimensions, it can be useful to think about two alternative approaches for counter-insurgency intervention, which this study calls "precautionary" and "remedial" counterinsurgency. These are ideal types: contrasting sets of related characteristics that illustrate the basic alternatives available to policymakers. As such, few if any real interventions will precisely match the models; instead, they should resemble one or the other to a degree that depends on where they fall along a continuum that the models describe.

Precautionary counterinsurgency is based on the idea that it is best to stop the growth of insurgencies early, before they develop into powerful armed movements capable of posing severe threats to local governments. Precautionary counterinsurgency interventions will thus tend to be relatively small overall and will tend to involve very limited military components. The military forces that are involved will normally focus on missions falling under the rubric of foreign internal defense, providing training, education, technical, and other assistance to local military and other security forces, although the bulk of the assistance most useful to states facing insurgencies in their earlier stages will usually be directed at improving policing, civil administration, and other nonmilitary functions. Examples of precautionary counterinsurgency often do not spring readily to mind, precisely because such interventions are small and are usually conducted without fanfare.[24] However, as the following chapters describe, the United States provides such assistance to a variety of governments facing insurgent threats around the world and is doing so with increasing frequency.

If intervening early against an insurgency can be equated to an ounce of prevention, *remedial counterinsurgency* is the pound of cure. In remedial counterinsurgency, the external power intervenes only when it becomes clear that the local government is not going to be able to suppress the insurgency on its own. The evidence for this is the successful growth of the insurgent movement into a powerful force, typi-

---

[24] Exceptions to this pattern most commonly occur when the intervention has relatively direct counterterrorist value, such as in U.S. security assistance to Georgia.

cally one that has wrested control of significant territory away from the government. Consequently, remedial counterinsurgency calls for large investments of resources to turn the tide, in particular for a larger military component than in precautionary counterinsurgency, and usually more-direct U.S. involvement in combat against the insurgents. The most prominent example of remedial counterinsurgency that the United States has undertaken was the intervention in South Vietnam in the 1960s.

As mentioned above, counterinsurgency interventions often involve elements of both ideal types and can be difficult to shoehorn into one category or the other. For example, the U.S. intervention in El Salvador in the 1980s involved a very small in-country military footprint, and U.S. advisers generally were not directly involved in combat against the Farabundo Marti Liberation Front (FMLN) rebels, yet the campaign, against a powerful insurgent movement, was a major effort of U.S. foreign policy to which considerable resources were devoted. However, the differences between the two approaches are very important for strategists to consider. In particular for military planners, precautionary and remedial counterinsurgency call for military forces with very different emphases: A force optimized for either type of counterinsurgency will be far from optimal for the other. The force optimized for precautionary counterinsurgency emphasizes small units conducting training and advising missions in peacetime or in more-limited combat situations. The force optimized for remedial counterinsurgency, while including training and advisory capabilities, emphasizes the ability to conduct large-scale operations across a large area.

A third type of counterinsurgency is worth considering as well, although it is in many ways better described as a variation on the second: conflict against an insurgency by an occupying power following a successful invasion or the complete collapse of a local government, as the United States has faced in Iraq since 2003, in Somalia from 1992 to 1993, and previously in the Philippines following the Spanish-American War.[25] For want of a better term, we refer to this as

---

[25] Invasions do not always lead to insurgencies. For example, U.S. operations in Grenada (1983) and in Panama (1989) did not face insurgencies during the occupation phase.

constabulary counterinsurgency." Like remedial counterinsurgency, this involves large commitments of forces, including combat troops; unlike it, however, precautionary counterinsurgency is not an available strategic alternative because no local government exists that would be capable of dealing with the problem even if it received suitable assistance. Instead, the occupying state should take what steps it can to limit the scale of the insurgency beforehand.

Advantages and Limitations of Precautionary Counterinsurgency

As implied by the ounce of prevention aphorism, and as the later chapters in this volume describe, a precautionary strategy seems to offer many advantages over the remedial approach. Interventions are dramatically smaller and less expensive in blood and treasure. The visibility of such interventions can be limited, thus minimizing the unfavorable political consequences for the host government that are likely when foreign powers are called in to help suppress domestic unrest. Yet precautionary counterinsurgency is not a panacea.

In deciding to focus on a precautionary approach, trade-offs are inevitable. Intervening early means intervening in many places instead of waiting to see which insurgencies will become serious problems if local governments are left to their own devices and then concentrating on those. In the words of Calvin Coolidge, "[i]f you see ten troubles coming down the road, you can be sure that nine will run into the ditch before they reach you." In practice, however, the costs of suppression escalate so steeply as an insurgency gains strength that, other things being equal, many precautionary interventions can be carried out for the price of a single remedial one. This makes precautionary counterinsurgency likely to be more efficient than remedial intervention, provided that the former has anything like a reasonable rate of success.[26]

However, even if the precautionary approach is consistently the most attractive choice, as this monograph argues, it is not one that can be relied on in every case. In at least four types of situations, early

---

[26] See Chapter Five for a comparison of costs in past U.S. counterinsurgency interventions.

intervention against an insurgency is either insufficient or simply infeasible. The first is that the insurgent threat may become visible only after it is too late for precautionary intervention. This could happen because intelligence sources or analyses overlooked the very existence of the insurgents until it was too late to intervene early. However, a far more likely scenario would involve recognizing the insurgents early on but not perceiving them to be a threat to the United States until after they had become powerful, either because they initially appeared to be a problem that the local government could contain or because the insurgents became hostile to the United States or its allies only after achieving considerable local success.[27] Even more simply, a second possibility is that the precautionary approach may be attempted, yet may fail to suppress or contain an insurgency, leaving the United States with a choice between giving up the effort and expanding it into a larger remedial intervention—although, if the precautionary effort was well executed and failed anyway, remedial intervention may not offer very attractive prospects for success.

A third possibility is that U.S. leaders will recognize the potential utility of precautionary action but choose not to carry it out because doing so would be politically costly for any of a variety of reasons, ranging from an isolationist electorate opposing intervention to support to an unsavory regime appearing to be politically or morally unacceptable. Finally, as already noted, some insurgencies are responses to foreign conquest or occupation, real or perceived; while it may be possible to prevent or limit the size of these resistance movements before they develop, the means for doing so will bear little resemblance to counterinsurgency operations.

Under any of these circumstances, having a preference for precautionary steps will not alter the facts if the choices have been reduced to remedial counterinsurgency and inaction. Doing nothing will usually be an option and may often be the best choice, but U.S. leaders will

---

[27] Intervening against insurgent groups that once appeared politically benign to U.S. interests but then produced a humanitarian catastrophe of unacceptable proportions would be similar. On the importance and challenges of recognizing dangerous insurgencies in time to intervene early, see Chapter Three and Bruce Hoffman, "Plan of Attack," *Atlantic Monthly*, July/August 2004b, pp. 42–43.

always want other options. Thus, even while emphasizing investment in the capabilities needed for a precautionary approach, U.S. strategists will not want to eschew remedial capabilities entirely.

Investing in Counterinsurgency Capabilities

Developing armed forces well suited for conducting counterinsurgency operations calls for significantly different patterns of investment than building forces optimized for conventional warfare. The expected balance between precautionary and remedial counterinsurgency will in turn have important implications for the specific form that this takes.

Optimizing Military Capabilities for Counterinsurgency

More than two years of U.S. counterinsurgency operations in Iraq have made lists of military capabilities that are particularly important in counterinsurgency a frequent and familiar part of contemporary defense policy discussions.[28] Therefore, the following treatment of this subject is deliberately brief, focusing on a small number of key points.

The first of these, already introduced earlier in this chapter, is that recasting military priorities to increase emphasis on counterinsurgency holds qualitatively different implications for the U.S. armed services. Counterinsurgency is primarily the business of land forces. This does not mean that air and naval power have little to contribute or that the Air Force and Navy do not need to make significant changes if the United States is to substantially increase its counterinsurgency capabilities. However, the extent to which air and naval forces can lighten the burden that falls on the ground forces is necessarily more limited here than in conventional warfare.

In the application of firepower, whether from air or surface forces, precision and discrimination are especially critical in counterinsurgency operations, in which the political costs of civilian casualties and collateral damage can easily outweigh the value of destroying intended targets. The intermingling of insurgents with civilian populations pres-

---

[28] The most authoritative treatment is found in U.S. Department of Defense (2006).

ents severe challenges for intelligence, surveillance, and reconnaissance capabilities and makes it desirable to have a variety of munitions with very limited effects. Concerns about the political side effects of military operations also usually make it attractive to be able to conduct them with a minimal visible presence of U.S. forces.

The ability of deployed U.S. forces to operate in contact with the local population without taking heavy losses is critical for a host of reasons. The intrinsic importance of keeping casualties low and the problem of maintaining domestic political support for prolonged operations entailing significant losses are obvious. It is also frequently important to avoid suffering casualties to avoid encouraging the insurgents with the not-unreasonable hope of driving U.S. forces away through attrition. Unfortunately, increasing the survivability of in-country forces is not just a matter of hardening their vehicles and installations, not least because this tends to interfere with the need for sustained interaction with the populace.

The enormous importance of information operations, civil affairs, intelligence, foreign area expertise, and associated capabilities in counterinsurgency, emphasized earlier in this chapter, need not be reiterated here. Other important functions range from providing medical services to placing advisers on defense policy and administration in host government ministries. Ground forces need not absorb the full burden of these missions. Indeed, the other services have much to offer in most of these areas and could expand their capabilities in others if necessary.

Beyond, and in many ways more important than, considerations relating to the need for changes in forces structure, counterinsurgency capabilities depend critically on what the priorities are in military education and training, including those for forces and personnel that do not specialize in counterinsurgency but would be expected to participate in it. This extends far beyond the need for relevant language and cultural sensitivity training and includes such challenges as routinely exercising counterinsurgency-like operations with joint, multinational, and interagency participation sufficient to live up to the slogan of "train like you fight."

When, How, and Where Will the United States Intervene?

Enumerating the sorts of capabilities counterinsurgency operations demand is only a first step toward designing appropriately sized and shaped forces to conduct the counterinsurgency missions of the future. Completing the architectural task will depend on expectations about future demands in at least three dimensions.

The most basic of these is the amount of counterinsurgency work likely to be called for, including the number, scale, and duration of expected operations. Sizing some capabilities will depend primarily on the total aggregate demand—the need for strategic airlift capacity to support operations is a simple example. For other capabilities, demand may be very different depending on whether 100 U.S. troops are to be deployed to each of 50 countries or 5,000 are expected to be sent to one place. For example, if a typical assistance package evolves to include one flight of unmanned aerial vehicles (UAVs), more of these units will be required to support a strategy of more small deployments than a few large ones.

The second issue is the expected balance between precautionary and remedial counterinsurgency operations, although this overlaps the quantitative dimension, since precautionary efforts should tend to involve considerably smaller forces overall. The approach to be emphasized shapes the types of forces required, with precautionary activities calling disproportionately for personnel to train and advise local government forces and remedial operations being more likely to require substantial direct support or combat activity involving U.S. forces. This choice also has profound implications for writers of counterinsurgency doctrine and for the sorts of arrangements that should be made and practiced for interagency integration in counterinsurgency operations.

Finally, there is the question—simple to ask but difficult to answer with confidence—of where future operations will occur. Predicting in which regions and even in which specific states future interventions are likely to be required is fundamental for developing the appropriate intelligence, linguistic, and local-area expertise before it is urgently required. Here again, embracing the precautionary model offers some benefits if it leads to a pattern of relatively steady rates of deployment to most parts of the developing world.

The answers to these questions will depend on U.S. leaders' fundamental choices about U.S. grand strategy, choices that in many cases have not yet clearly been made as of the fifth year of the global war on terrorism. They are tied to the issue of when insurgencies pose threats to U.S. interests serious enough to merit military intervention on either a small or a large scale, as discussed at the beginning of this chapter. Predicting how willing or unwilling Washington will be to accept the replacement of friendly but ineffective governments by more hostile challengers, the creation of power vacuums as a result of state failure or loss of territorial control, or the apparent success of governmental usurpers in general or Islamist ones in particular is a task that falls beyond the scope of this analysis. However, this will be the central determinant of both the size and the shape of the demands for counterinsurgency capabilities that military leaders will need to prepare to provide.

The Roles of Allies in Counterinsurgency

Counterinsurgency operations that U.S. armed forces conduct are by nature multinational. Not only do they occur in the territory of other states,[29] since the United States is not a colonial power like France in Algeria or Britain in Malaya, but defeating an insurgency is ultimately a function of the relationship between the local government and its people—and where no government exists, one must be created. This means that, even when U.S. forces do the bulk of the fighting, they are not fighting by themselves, and victory cannot be achieved by U.S. efforts alone, especially since the labor-intensive nature of counterinsurgency makes reliance on local manpower essential even aside from the need for linguistic and local-area expertise.[30] Much has already

[29] It is possible to conceive of operations against certain U.S. militant groups as counterinsurgency, but these fall into the domain of domestic law enforcement rather than national security policy.

[30] On both the necessity and the challenges of collaborating with local governments facing insurgent threats, see Daniel Byman, *Going to War with the Allies You Have: Allies, Counter-*

been said in this study about the enormous importance of the relationship between the local government and its military, police, and other forces on the one hand and those of the United States on the other.

U.S. and local forces may not be the only ones involved in counterinsurgency campaigns, however, and this chapter's discussion of grand strategy would not be complete without devoting some attention to the potential roles of third-party allies and coalition partners in such operations alongside the United States.

The U.S.-led counterinsurgency campaign in Iraq since 2003 has been a significantly multinational effort; by September 2005, the armed forces of 14 nations in addition to the United States and the United Kingdom had suffered fatalities in the fight against Iraqi insurgents. This is not a new development—although Americans rarely recall it, the U.S. intervention in Vietnam saw the forces of several allied states, including Australia, New Zealand, and South Korea, fight alongside U.S. and South Vietnamese troops. Even token contributions of forces or other support can strengthen U.S. efforts by helping to legitimize the intervention. However, even more than in conventional warfare, allied participation in counterinsurgency operations can make important military differences.

There are several reasons for this. Because, as discussed above, counterinsurgency tends to be very labor intensive over prolonged periods, it is often difficult to muster and sustain optimal levels of manpower for large remedial interventions, and allied forces can help with this challenge. More important, allies may be able to contribute types of forces that are essential but consistently in short supply even for precautionary actions, such as special operations forces (SOF).[31] Even

---

insurgency, and the War on Terrorism, Carlisle Barracks, Pa.: Strategic Studies Institute, U.S. Army War College, November 2005b.

[31] During Operation Enduring Freedom, a number of NATO allies quietly contributed SOF, especially ones with mountain warfare experience, to augment U.S. SOF deployments. Similarly, during the initial phases of OIF, SOF were the most significant contributions of Australia and Poland, the only countries to join in the invasion along with American and British forces. See Nora Bensahel, *The Counterterror Coalitions: Cooperation with Europe, NATO, and the European Union*, Santa Monica, Calif.: RAND Corporation, MR-1746-AF, 2003.

more relevant to counterinsurgency, with its demands for conducting policing under conditions of threat much greater than those faced by American civilian law enforcement agencies, are paramilitary police forces, such as Spain's *Guardia Civil*, Italy's *Carabinieri*, and France's *Compagnies Républicaines de Sécurité*. The United States has no equivalent organization capable of sharing its expertise with local government forces that need to develop similar capabilities, so U.S. allies could be invaluable in filling this gap.

Depending on the location of the intervention, certain allies (such as former colonial powers) may possess area expertise, specialized operational experience, or local connections that can be extremely valuable to a counterinsurgency campaign. Finally, substituting non-U.S. for U.S. forces may help moderate political backlash against foreign intervention when a visible U.S. presence is particularly incendiary; in Arab or other predominantly Muslim states, troops from other Muslim countries could potentially be especially helpful in this respect. For all these reasons, if the United States is going to be interested in doing a considerable amount of counterinsurgency intervention, it would be well served to seek out allied participation when possible and even be open to the possibility of playing a supporting role in combined interventions in which non-U.S. forces take the lead.

Taking advantage of these potential contributions from allied states requires not only interoperability in technical areas, such as communications systems, but also compatible doctrines and philosophies about the strategy and tactics of conducting counterinsurgency. Expecting other armed forces simply to adopt U.S. doctrine would be naïve—it would also likely be dysfunctional when dealing with states that in some cases have experience in counterinsurgency that rivals, and arguably surpasses, that of the United States.

Is it realistic to expect other countries to play such an important part in intervening against insurgents who threaten U.S. interests? Among the Western allies, this is what might be called a historically natural role only for the United Kingdom, France, and Australia. However, the traditional security threats against which European states have long focused their attention continue to recede, leaving their armed forces increasingly in need of a mission,[32] and to a significant degree Islamist militants threaten not only U.S. interests but those of the West more generally, as is reflected in the considerable European contributions of forces to stabilization operations in Afghanistan since the fall of the Taliban. Moreover, the levels of forces and resources required to provide significant contributions to precautionary counterinsurgency operations are easily within the reach of even third-tier Western powers, and such activities ought to be palatable even to populations that are reluctant to see their armed forces engaging in combat operations far from home. Therefore, it does not appear unreasonable to envision the security policies and military capabilities of other U.S. allies, or of the North Atlantic Treaty Organization (NATO) as a whole,[33] becoming increasingly focused on counterinsurgency operations in coming years, particularly if the United States offers appropriate encouragement and support for such a trend.

Having considered counterinsurgency from the perspective of U.S. grand strategy, we next present a conceptual framework that more fully develops the rationale for and capabilities associated with a precautionary counterinsurgency strategy.

---

[32] Traditional peacekeeping operations are arguably less likely to provide this occupational niche to a fully satisfying degree over the long term than they once appeared to, especially as the Balkans gradually stabilize.

[33] Paul Ames, "U.S. Urges NATO to Take Role of Trainer," *Washington Post*, September 22, 2005.

CHAPTER FIVE
A New Framework for Understanding and Responding to Insurgencies

This chapter presents a conceptual framework to guide our thinking about the problem of insurgency, the relative advantages of various intervention options, and the most effective use of military power. Arguing for a precautionary strategy that would seek to head off potential insurgencies before they reach a critical mass, the chapter then explores the role of security cooperation, using the U.S. assistance to the government of El Salvador in the 1980s to illustrate both the power and limitations of military assistance.

The Application of Military Power to Counterinsurgency

As discussed in Chapter Two, insurgency and counterinsurgency are primarily political in nature. Police, civilian security, and intelligence forces are the state's most important coercive instruments in defeating an insurgency. Military power plays a secondary and supporting role in counterinsurgency. The military instruments of a state are nevertheless critical participants in many counterinsurgency campaigns, primarily as a buttress to domestic security services, and the appropriate application of military power is a central challenge for policymakers seeking to define an effective counterinsurgency strategy.

Since the late 19th century, the United States has participated in counterinsurgency campaigns as an external power, seeking to assist a partner government in its struggle against an insurgent threat.[1] This external role creates challenges for the United States as it seeks to apply its various instruments of national power to affect a political conflict within a society and culture that are naturally alien. U.S. intervention can bring important resources and capabilities to the counterinsurgency effort, but it also risks undermining the legitimacy of the partner government, stoking the resistance, and creating dependency relationships in the host society and government. This is particularly likely when U.S. military power is committed to a foreign counterinsurgency campaign. U.S. policymakers are therefore confronted by the need to shape and meter U.S. military involvement to ensure, to the extent possible, that its contributions are greater than its liabilities.

The United States possesses a variety of instruments and capabilities applicable to the counterinsurgency problem, ranging from indirect economic, political, intelligence, and military assistance programs to the commitment of U.S. combat forces. As U.S. policymakers consider involving the nation in a counterinsurgency campaign, their central challenge is to weave the available instruments into an appropriate strategy that supports the legitimacy and capacity of the host-nation government. Where the military instrument is concerned, policymakers will confront three key questions as they develop an appropriate strategy. The first is the timing of U.S. military involvement, whether to commit military resources early in a developing insurgency, while the threat is nascent and unpredictable, or to wait for insurgent threats to develop more momentum and a definitive shape before choosing to commit U.S. military resources. The second question is the nature of U.S. military involvement, whether to focus on indirectly assisting and advising local security forces or on directly committing U.S.

---

[1] Many consider the 19th century "Indian Wars" to be the most recent counterinsurgencies on U.S. soil. They concluded in 1890 at Wounded Knee. If colonial territories are considered, the Philippine Insurrection (1899–1913) is the best candidate for the most recent counterinsurgency campaign on U.S. soil. Elsewhere, the United States has supported insurgents against existing governments in a few cases. The most obvious examples are Afghan insurgents during the Soviet occupation in the 1980s and then again in 2001, "Contra" rebels in Nicaragua in the 1980s, National Union for the Total Independence of Angola insurgents in Angola (1980s), Mozambican National Resistance insurgents in Mozambique (1980s), and the Sudan People's Liberation Army in the 1990s.

forces to operations against the insurgents. The third question is the degree of U.S. military profile or "footprint" appropriate to the local political circumstances and the overarching counterinsurgency strategy pursued by the host-nation government and the United States. These three issues are interrelated in complex and important ways. To explore and better understand these complex interactions, it is helpful to think of an insurgency as a dynamic phenomenon that grows or shrinks in response to its environment.[2]

Figure 5.1 illustrates a simple insurgency that enjoys increasing success over time. The curves that branch away from the main curve illustrate the potential effects of intervening at different points. In the

Figure 5.1
Intervention Options Against a Growing Insurgency

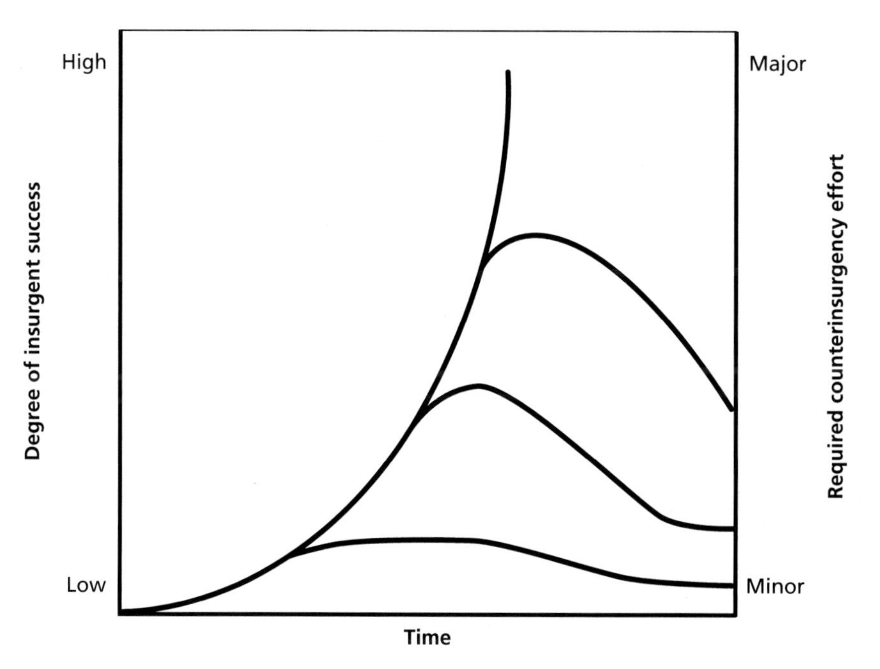

RAND *MG509-5.1*

[2] David Kilcullen, "Countering Global Insurgency," *Small Wars Journal*, November 2004.

ideal situation, good government, a fair and just society, and economic opportunities would have prevented an insurgency from arising in the first place. Even these strong societies are likely to have their individual malcontents but without some underlying grievances to engage a larger audience, an insurgency will not gain traction. When there is more-fertile soil for insurgents to work with, good intelligence and police work can stop the insurgency at an early stage, as shown by the lowest branching curve (although political, economic and other reforms are still likely to be necessary to address the underlying roots of the insurgency). Over time, however, the risks and costs of countering the insurgency increase dramatically. The middle and highest branching curves illustrate the greater effort required and greater risk associated with later interventions. From a U.S. perspective, the objective of counterinsurgency strategy is to reverse the momentum of a threatening insurgent movement as quickly as possible and with the fewest U.S. military resources as possible.

In the preventive category of U.S. military involvement, we include basic professionalization, individual training, and education provided most often through security assistance channels, with a prime example being the International Military Education and Training (IMET) program.[3] The essential aim of preventive involvement is to increase the basic functioning and capacity of the partner nation's military institutions. It also can, in theory at least, expose partner-nation officers to key ideas supporting the long-term viability of legitimate government, including democracy, primacy of civilian authority, and respect for human rights. By increasing the professionalism of the military institutions and, potentially, their acceptance of the key pillars of legitimate government, preventive involvement ideally helps make the partner nation less prone to the development of a domestic insurgency and more capable of suppressing a nascent insurgency within acceptable bounds. Furthermore, basic professionalism provides the essential

---

[3] Other tools of preventive involvement are mobile training teams (MTTs), seminars, conferences, individual training funded by Foreign Military Financing, familiarization exchanges, and many others. DoD refers to these instruments collectively as *security cooperation.*

foundation for any subsequent training and assistance the U.S. government may wish to provide to host-nation military forces. Absent a basic level of professionalism, host-nation forces are all too likely to employ their growing capabilities in ways that are counterproductive to the overall effort.[4]

Once the insurgency has become established but is still relatively young and small, the counterinsurgent's focus shifts from prevention to direct action against the rebels, their strategy, and their sources of support. For the U.S. military, the primary roles at this stage include collective training of partner-nation military units, the provision of equipment and support items, the focused and coordinated provision of both equipment and collective training in key mission areas, and long-term training conducted by U.S. personnel embedded in the partner military.[5] The central objective of this type of assistance, and what distinguishes it from purely preventive engagement, is that these activities are aimed at cultivating specific capabilities within partner militaries.[6] These capabilities must be carefully tailored to account for the nature of each partner's particular counterinsurgent challenge, the domestic political situation, and the relative strengths and weaknesses of the host-nation armed forces. The objective is certainly not to create a mirror image of the U.S. armed forces but rather to provide the discrete capabilities that will make the host-nation armed forces as effective as possible against the unique threat they face.

---

[4] It is vital to note that professionalization alone does not guarantee that a military will be competent at counterinsurgency. Professionalism primarily addresses the role of the military in society and secondarily its fundamental military skills. A professional military may or may not be skilled at fighting insurgents.

[5] Embedded training is not common today for the U.S. military, but some allied militaries (including, most prominently, the British) employ embedded trainers extensively.

[6] There will also be a crucial requirement for assistance to host-nation law enforcement, security, and intelligence services. We believe DoD is poorly suited to this mission, quite apart from the prohibitions in Section 660 of the Foreign Assistance Act. An implicit requirement for a precautionary approach to counterinsurgency may therefore be the expansion of the capacities of U.S. government civilian agencies to assist foreign law enforcement, security, and intelligence services.

This category of involvement will tend to involve higher-risk activities than would purely preventive involvement, including, for example, the deployment of some U.S. personnel to partner nations. These activities will also tend to be more expensive than preventive activities, and the U.S. profile will be more prominent as a result. Depending on the circumstances, these may be important liabilities, and the decision to undertake this greater involvement in a foreign insurgency must balance the potential benefits of these activities against the cost and broader ramifications of greater U.S. involvement.

If the United States waited until the insurgency had worsened or if it grew despite earlier support to the partner nation, more-direct U.S. involvement might be warranted. This could include advisory missions, meaning direct U.S. military assistance in preparing for and, perhaps, planning specific military operations and noncombat support from U.S. operational units. This can range from *noncombat advising*, which helps a partner military plan operations that do not involve combat, through *combat advising*, which involves helping partners plan operations requiring lethal force, to *embedded advising*, which places a U.S. officer within a partner military organization on a long-term basis to serve as a full-fledged member of that unit, including advisory activities during combat and noncombat operations. The involvement may also include operational units of the U.S. armed forces conducting noncombat support operations in theater, including intratheater and vertical lift, intelligence, surveillance, reconnaissance, communications support, and a variety of other key support missions. The central aim at this point in the conflict is to cultivate a broad set of key capabilities within a partner military and sustain them over a significant period. A secondary aim might be to provide direct noncombat support to key partner-nation military operations. The move to these more-invasive approaches to counterinsurgency is therefore marked by an expansion in the scope of assistance provided and the time frame over which the U.S. expects to be involved. This level of involvement in counterinsurgency will tend to be more expensive and riskier than preventive or limited indirect assistance activities, both for the United States as a whole and for the individual military units conducting the activities with the partner nation.

In the final category of counterinsurgency involvement (akin to intensive care for a seriously ill patient), we include direct U.S. combat operations in support of partner-nation forces or on their own accord. In these instances, as in the medical world, critical functions have been undertaken by outsiders. Direct U.S. combat operations will tend to be the most risky and expensive alternative for the United States, but they may nevertheless be necessary in cases of dire emergency.

The relative risk and expense of the various counterinsurgency options will be key considerations for U.S. policymakers choosing among categories of counterinsurgency involvement. The various intervention points identified in Figure 5.2 depict historical examples that we believe illustrate the trade-offs between time, resources, and risk.

In the lower left corner of Figure 5.2, at the earliest point of involvement, we identify International Military Education and Training as a preventive activity. It can be conducted very early, creates a very low American profile (most courses are held in the United States), and is relatively inexpensive. The particular example cited here involved sending a foreign student to the Army's Basic Noncommissioned Officer Course (BNCOC), which costs roughly $5,500 per student.[7] Many thousands of such students are educated in U.S. military schools every year, supplemented in some cases by Mobile Education Teams visiting partner nations from the United States.

A higher level of involvement occurs later and in the context of an insurgency posing a greater risk. The two illustrative examples identified in Figure 5.2 are a Joint Combined Exchange Training (JCET) exercise and the Georgia Train and Equip Program (GTEP). JCETs are typically conducted by U.S. SOF working in conjunction with partner-nation forces for, in most cases, a month or slightly longer. Legally, the primary purpose of the JCET is to allow U.S. forces to practice advising foreign forces. The secondary, though in practice no less essential, purpose of JCETs is to allow U.S. forces to cultivate a particular set of capabilities in partner militaries. The costs of JCETs are relatively low,

---

[7] Data extracted from U.S. Department of State, "Foreign Military Training and DoD Engagement Activities of Interest," Washington, D.C., 2005.

Figure 5.2
Examples of Counterinsurgency Intervention Strategies

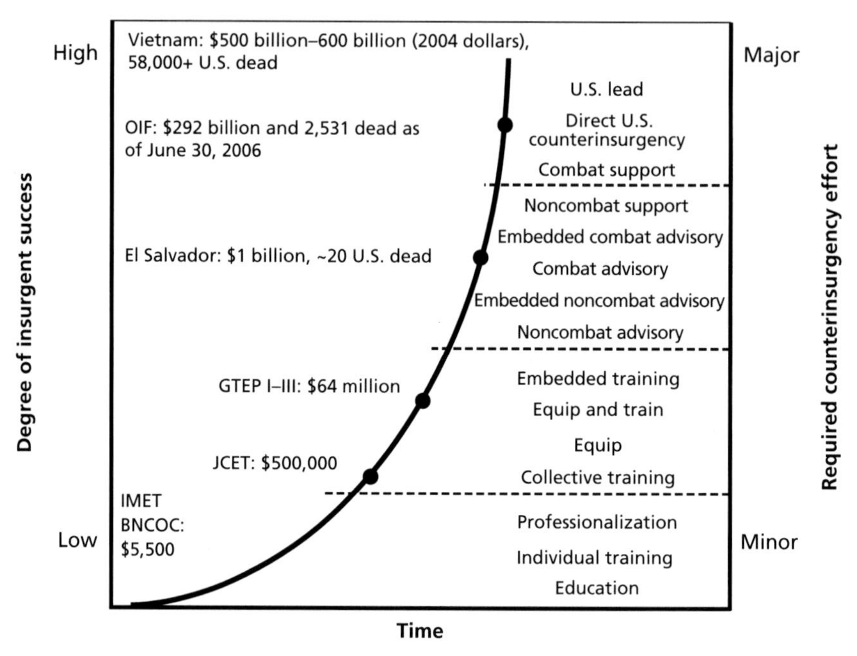

RAND *MG509-5.2*

approximately $500,000, depending on the location of the exercise, the objectives of the training, and the type of team the United States provides. The U.S. military typically conducts dozens of JCETs per year.[8]

GTEP aimed to provide a significant counterterrorism and counterinsurgency capability to the armed forces of the Republic of Georgia. The government of Georgia has in recent years found it difficult to extend its control over the Pankisi Gorge region of the country, where there are armed groups of ethnic Chechens operating in relative impunity. The first three phases of the program provided individual and collective training for four battalions of the Georgian Army and a consignment of new or refurbished equipment including UH-1 Iroquois helicopters. The cost to the U.S. government of the first three phases of the GTEP program was $64 million. Despite early chal-

---

[8] Interviews with theater SOF component planners, November 2004.

lenges, the results of the GTEP program are promising.[9] The Chechen groups have departed the Pankisi Gorge region, Georgian forces appear to be increasingly capable of securing the republic's territory, and the Georgian government has even contributed its newly trained battalions to OIF.[10] The GTEP program has been extended in subsequent years to incorporate a greater proportion of Georgian forces. The JCET and GTEP examples suggest how more-active involvement differs from purely preventive involvement, involving higher cost and risk for the potential opportunity to cultivate a specific capability. Of course, programs like JCETs and GTEP can and should be employed in conjunction with preventive approaches, such as professionalization and IMET programs. These simultaneously help create a productive environment for larger-scale activities while also helping ensure that the benefits of these activities will be sustained over time by a professionalized and Western-trained officer corps.

If the United States waits until an insurgency has progressed beyond these early stages or if more-limited approaches fail, it may be necessary to employ an invasive approach. Here, we are referring to long-term, broad-based assistance programs that are coupled with formal or informal advisory assistance and, perhaps, noncombat support operations by U.S. units. The illustrative example provided in this category is the U.S. assistance effort to El Salvador in the 1980s, an involvement that was hugely controversial at the time and dogged by many setbacks but that was ultimately successful, as described later in this chapter. The U.S. involvement in El Salvador included training nearly the entire officer corps of the Salvadoran armed forces, provision of massive material assistance, collective training for most Salvadoran operational units, and noncombat support from U.S. aircraft and ships.

---

[9] U.S. European Command, Georgia Train and Equip Fact Sheet, undated [2003]. See also John D. Banusiewicz, "Rumsfeld Visits Georgia, Affirms U.S. Wish That Russia Honor Istanbul Accords," *American Forces Press*, December 5, 2003.

[10] See Elman Agayev, Mamuka Kudava, and Ashot Voskanian, "U.S. Security and Military Cooperation with the Countries of the South Caucasus: Successes and Shortcomings," event summary, Harvard Belfer Center for Science and International Affairs, Caspian Studies Program, May 13, 2003. It should also be noted, however, that some observers do not believe that GTEP played a decisive role in the departure of the Chechens from Pankisi.

Many at the time considered this involvement, which topped $4.5 billion in economic aid and over $1 billion in military aid and cost the lives of 21 U.S. personnel, to be profligate.[11] As in other cases, the United States did not pursue invasive counterinsurgency techniques in isolation but rather built on a foundation of preventive and other activities both to prepare the groundwork for more-intensive U.S. assistance and to ensure, insofar as possible, that the capabilities the United States provided would be sustained in the context of a professionalized and effective Salvadoran military.

Finally, if the United States waits until an insurgency situation is truly dire or if previous efforts fail to stem the tide, it may be necessary for U.S. forces to take a direct combat role against insurgents. This is, of course, tremendously risky and costly. The two illustrative examples provided in Figure 5.2 are the ongoing counterinsurgency being conducted as part of OIF and the Vietnam War. OIF is a useful illustration of the potential scale and scope of U.S. involvement, rather than an illustration of how the United States will likely find itself involved in counterinsurgency in the future (that is, as an occupying power of a major regional state). The counterinsurgency component of OIF has consumed the better portion of U.S. military power in Iraq since mid-2003, and the Congressional Budget Office estimates that the operation will eventually cost between $273 and $392 billion, in addition to thousands of dead and wounded.[12] Vietnam is another cautionary tale, one in which the United States tried but failed to limit its involvement to advising, training, and equipping. The combat phase of the war in Vietnam eventually cost more than $500 billion in fiscal year (FY) 2004 dollars and, far more importantly, 58,000 dead and many thousands more wounded, injured, and missing.[13]

---

[11] Corum and Johnson (2003, p. 327) and Bradley Graham, "Public Honors for Secret Combat," *The Washington Post*, May 6, 1996, p. A1.

[12] Congressional Budget Office, *Estimated Costs of Continuing Operations in Iraq and Other Operations of the Global War on Terrorism*, Washington, D.C., June 25, 2004.

[13] Care must be taken in using the Vietnam War as an example of counterinsurgency costs and risks. The war was in part an insurgency, in part a civil war, and in part an international conflict waged in the two Vietnams, Laos, Cambodia, and Thailand. Ultimately, the outcome was decided when some 17 North Vietnamese Army divisions invaded the South and

The key insight from this analysis is that the costs of an intervention skyrocket when U.S. operational units become involved. The comparison between El Salvador and OIF is most instructive in this regard. Although considered expensive at the time, the American involvement in El Salvador was miniscule compared to our commitment to Iraq. In fact, the United States could conduct one El Salvador–level effort (e.g., over $1 billion in direct military assistance over most of a decade[14]) in every nation of the world for far less than the cost of an OIF-level involvement in some future counterinsurgency.[15] Likewise, the United States could conduct 16 interventions at the level of GTEP ($64 million) for the cost of a single El Salvador–type effort. Stated another way, as of June 2006, OIF has already cost 4,500 times as much as the GTEP program.

These relative costs can also be stated in terms of probabilities. As noted in Chapter Three, one of the great weaknesses of the U.S. counterinsurgency capacity is our lack of an effective indications and warning system for predicting which insurgencies are likely to become dangerous in the future. While true, the cost differential between the types of counterinsurgency involvement suggests that low-level preventive counterinsurgency techniques are cost-effective even when it is extremely unlikely that any given activity or intervention will yield a

---

reunified the country via conventional combat. (Our thanks to RAND colleague Stephen Hosmer for making this point.) War financial costs from Andrew Krepinevich, *Iraq and Vietnam: Déjà Vu All Over Again?* Washington, D.C.: Center for Strategic and Budgetary Assessments, July 8, 2004, pp. 11–12; war casualties from U.S. Army Center for Military History, "Vietnam Conflict—Casualty Summary," Web page, June 15, 2004.

[14] Estimates for the total cost of vary. Schwarz (1991, p. 2) estimates the total of direct military and economic aid to be at least $4.5 billion, over $6 billion if Central Intelligence Agency activities and unsubsidized credits are included. Philip J. Williams and Knut Walter, *Militarization and Demilitarization in El Salvador's Transition to Democracy*, Pittsburgh, Pa.: University of Pittsburgh Press, 1997, p. 133, presents data from U.S. Agency for International Development and Congressional Research Service documents that suggest direct military aid for the years 1980 to 1990 was slightly over $1 billion. We used the latter figure for this comparison.

[15] There are 191 member states in the United Nations, and multiplying that by 1 yields $191 billion. OIF costs as of June 30, 2006, were $292 billion. The Congressional Budget Office estimates total costs could approach $400 billion. See United Nations, "List of Member States," Web page, February 10, 2005.

decisive result. If it is true that the United States could conduct nearly 200 interventions on the scale of El Salvador for less than an OIF-level operation,[16] it is also true that, if an El Salvador–scale intervention has even a 0.5-percent chance of averting an eventual OIF-level American involvement, it would be cost-effective to conduct such an intervention. Likewise, if OIF is 4,000 times as expensive as GTEP, a GTEP-scale effort that had even a 0.025-percent chance of averting an OIF-level effort would be beneficial.

Although these cost comparisons are in some ways unrealistic—they do not reflect the how U.S. policy is actually made—they underscore the fundamental point that it is preferable for the United States to involve its military instruments as early as possible and in as small and discrete a manner as possible. Even when it is unclear whether an insurgent threat is likely to emerge, it will make sense to become involved in preventive measures if there is the slightest possibility that a large-scale commitment of U.S. forces might thereby be averted.

Within USAF, the pilot-retention bonus program operates on an analogous principle. The bonus is offered to any pilot who signs on for an additional period. Some of those who receive the bonus would have stayed on without it. USAF cannot know who those pilots are, so it pays bonuses to all who stay. For a relatively small investment, USAF keeps experienced pilots in the force and saves the cost of bringing an aviator up to the ten-year experience level. The program works as follows:

> As with any retention bonus, the marginal cost of an additional year of retention is not the cost paid to a single pilot. Rather, it's the cost paid to all pilots divided by the marginal increase in those who stay. For example, there are ten pilots approaching the 11th year of service and you offer each of them $20,000 to stay an additional year. Six of them would have stayed without the bonus. The cost for each additional manyear is $200,000 divided by 4, or $50,000.[17]

---

[16] This is not meant to suggest that the United States will be conducting 200 simultaneous assistance missions but rather to show the relative costs.

[17] Personal communication with Al Robbert, Director, Manpower Personnel and Training Program, Project AIR FORCE, RAND Corporation, July 6, 2005.

Similarly, some insurgencies will go away on their own or because the local government acts effectively against them. Others will grow into something dangerous without outside support to the host nation. Since we cannot consistently predict the growth pattern of insurgencies, the precautionary strategy—like the pilot bonus program—seeks to spend a modest amount in many places to ensure that assistance is available in those situations where it is required.

Security Cooperation and Foreign Internal Defense

As we argue above, a precautionary strategy calls for greatly expanded security cooperation between the United States and its partners and allies around the world.[18] Security cooperation can contribute to foreign internal defense in several ways. Two stand out in particular: professionalization of friendly militaries and expansion of their counterinsurgent capabilities. Where professionalization is concerned, security cooperation can help new nations or those taking first steps toward democracy establish the preconditions for a successful transition to democratic rule. For example, security cooperation programs have helped inculcate democratic values among commissioned and noncommissioned officers (NCOs) in military forces of many nations.[19]

---

[18] *Security cooperation* is the term DoD uses to denote "those activities conducted with allies and friends, in accordance with Secretary of Defense Guidance, that: Build relationships that promote specific U.S. interests; Build allied and friendly capabilities for self-defense and coalition operations; [or] Provide U.S. forces with peacetime and contingency access." These activities include "Combined Exercises, Security Assistance, Combined Training, Combined Education, Combined Experimentation, Defense and Military Contacts, Humanitarian Assistance, and OSD-managed programs." The annual Department of Defense Security Cooperation Guidance is classified. The unclassified excerpts above were provided in a briefing by then–Deputy Assistant Secretary of Defense Andrew Hoehn in April 2003. For a more-detailed discussion of DoD security cooperation planning, see Thomas Szayna, Adam Grissom, Jefferson Marquis, Thomas-Durell Young, Brian Rosen, and Una Huh, *U.S. Army Security Cooperation: Toward Improved Planning and Management*, Santa Monica, Calif.: RAND Corporation, MG-165-A, 2004, especially Chapter Two, "The Security Cooperation Planning Process: Its Evolution and Current State."

[19] The Expanded IMET program is particularly important in this regard, as are the DoD regional centers (the George C. Marshall European Center for Security Studies, the Near

Although having a professional military that observes the rule of law will not necessarily prevent insurgencies from arising, a military that is violent and oppressive creates a legitimate grievance that could lead to violent rebellion.

Where counterinsurgency capabilities are concerned, security cooperation allows the United States to cultivate tactical and operational competence in foreign militaries facing insurgent challenges. The subject matter may range from very low-level tactical capabilities, such as effective combined-arms patrolling, to sophisticated topics, such as the art of interweaving of military and civil actions at the operational level.[20] A security assistance effort with an ally or partner may include a suite of cooperative activities ranging from individual training at U.S. schoolhouses, through materiel grants and sales, to direct advisory assistance. In some cases, security cooperation yields quick results, but in most cases, it requires patient and thoughtful DoD effort over years or decades.

At its most effective, in the Foreign Internal Defense context, security cooperation melds professionalization and capacity-building to cultivate local military forces that are more capable, but also more tactful, in pursuing insurgents. Employed in the right circumstances, security cooperation, along with economic and political aid and assistance, can help partner nations develop good governance and the military competence to head off insurgencies or defeat those already in place.

East South Asia Center for Strategic Studies, the Asia-Pacific Center for Security Studies, and the Center for Hemispheric Defense Studies). For perspective on how this worked in one region, see Graeme Herd and Jennifer Moroney, eds., *Security Dynamics in the Former Soviet Bloc*, London, UK: Curzon, 2003.

[20] As we noted earlier, the U.S. military has a relatively small number of personnel with the counterinsurgency experience and expertise to advise and train on the more-sophisticated aspects of counterinsurgency doctrine and operations. A precautionary strategy along the lines envisioned here would require a significant increase in the number of military personnel with counterinsurgency expertise and likely would also necessitate the expansion of organizations (e.g., Army Special Forces groups) that specialize in training and advising partner-nation personnel.

Security cooperation has been an important part of U.S. foreign and defense policy for many decades.[21] The United States has used it to provide material support to allies during wars, to support diplomacy, and to help friendly nations improve their ability to defend against internal and external foes. The United States provides military aid to nations in every region of the globe, funding roughly $4.6 billion in Foreign Military Financing (FMF) and $91 million in IMET in FY 2004. Yet 90 percent of the FMF funds go to just six nations, leaving roughly $500 million for the rest of the world.[22] Although these programs have been effective in achieving U.S. objectives in the past, they may need to adapt to new circumstances and goals.

The threat of global terrorism has led the U.S. government to increasingly emphasize what it calls "capacity building" to help foreign partners develop strong counterterrorism capabilities. Although U.S. Central Command (CENTCOM) efforts in Afghanistan and Iraq dominate the news, European and Central Commands have significantly increased such efforts in Africa, as has Pacific Command in Southeast Asia. These efforts are commendable but are often constrained by security assistance legislation, funding mechanisms and programs that are holdovers from the Cold War, and a lack of sufficient resident security cooperation capacity within the U.S. military departments.

To defend itself effectively against an evolving global terrorist and insurgent threat, the United States will need flexible and adaptable policy instruments. In particular, previous distinctions between peacetime and wartime activities have less and less meaning in a world in which terrorists can strike anyplace at any time.

---

[21] Modern DoD security cooperation is an outgrowth of activities begun during the first counterinsurgency era under the auspices of the U.S. Foreign Assistance Act of 1961 (and amendments), the Arms Export Control Act (and amendments), and related statutes.

[22] In FY 2004, the top six recipients of military aid were Israel ($2.1 billion), Egypt ($1.3 billion), Afghanistan ($364 million), Jordan ($204 million), Colombia ($98 million), and Pakistan ($75 million). See U.S. Department of State, "International Affairs Budget," Web page, Washington, D.C., 2006, and U.S. Department of State, Office of Plans, Policy, and Analysis, "International Military Education and Training Account Summaries," Web page, Washington, D.C., 2006.

During the Cold War, most advisory and training activities were constrained by "peacetime" restrictions. Training deployments were typically limited to a few weeks or months; advisors were generally not allowed to accompany their units into combat; advisory organizations were underfunded and undermanned; and career paths were often unclear or unattractive. In the new security environment, these activities need to be given the highest priority, well-qualified personnel, and significant funding. Security cooperation legislation, organizations, funding, programs, and concepts all need to be revisited and updated in light of new security challenges. In particular, U.S. capabilities, funding, and programs for Foreign Internal Defense are not well integrated at the national level.

MTTs, which visit host nations for a few weeks or months, will continue to be the instrument with the widest applicability.[23] They provide excellent opportunities to build the capacity of friendly armed forces to conduct counterinsurgency operations. MTTs can be tightly focused on a partner's most pressing needs, scaled up or down to match the absorptive capacity of the partner and the vicissitudes of bilateral political-military relationships, and synchronized with other security cooperation instruments (such as FMF grants) to make the most of limited U.S. resources. Experience and deployment data suggest, however, that roughly three visits annually are generally required to build more-sophisticated capabilities and to have lasting effects. At present, the limited MTT capacity of the U.S. armed forces makes it impossible to interact at this high level with the large number of potential partners. One option to supplement these MTTs is to embed advisors in partner nations for 1- to 2-year tours, an approach the British have used successfully. The British Ministry of Defence has for decades embedded commissioned officers and NCOs in a number of foreign militaries in Africa, the Middle East, and Southeast Asia. These officers fulfill a number of roles, ranging from direct command

[23] Training can be, and has been, conducted under a number of different auspices, including mobile education teams, JCET exercises, State Partnership Program events, Joint Contact Team Program events, and many others. We use the term "mobile training team" to describe all these types of events. The MTT label both is widely used and captures the essential nature of the interaction.

of partner-nation units, through advisory posts, to professorships in foreign command and staff colleges. One recent example is the International Military and Advisory Training Team (IMATT) established in Sierra Leone in 2002 to expand the capacity of the Republic of Sierra Leone Armed Forces (RSLAF) to control the nation's territory and suppress warlords and bandits. British officers both advise their Sierra Leonean counterparts and directly command some elements of the RSLAF. IMATT provides a brigade advisory and support team to each RSLAF brigade, trains RSLAF soldiers at the Benguema training center, and directly administers the RSLAF officer academy. British IMATT personnel are drawn from all three services, and their specialties range across the combat, combat support, and combat service support functions. IMATT has been widely credited with suppressing warlordism in Sierra Leone and contributing to stability in west Africa more broadly. Recently, IMATT has been augmented by officers from Bermuda, Canada, Ghana, Jamaica, Nigeria, Senegal, and even the United States.[24]

Recent U.S. experience in Iraq and Afghanistan has confirmed the great value of embedding advisors. Embedded advisors gain an insider's understanding of the strengths and weaknesses of the partner military and share hardships and risks with the host-nation forces over several years. Consequently, the advisors develop lasting relationships and credibility with the locals. It is therefore notable that the U.S. armed forces possess no institutional mechanism for embedding commissioned officers or NCOs in foreign militaries.

U.S. Army Special Forces deploy the largest number of advisors and trainers today. Counterinsurgency operations require well-trained and disciplined ground forces, and U.S. Army advisors are likely to be in increased demand as the United States increases emphasis on counterterror and counterinsurgency operations. Since U.S. Army SOF

---

[24] Conversations with UK Ministry of Defence policy staff, February 2005. See also The International Military Advisory and Training Team (Sierra Leone), Web site, February 8, 2005.

capabilities and contributions are widely appreciated and understood, we will not go into a detailed discussion of their mission here.[25]

In the past, U.S. training and advising have focused on partner nations' land forces. Air and naval operations were often neglected because the recipient nation lacked an air force or navy or because they were limited in their capabilities. Although the specific requirements will vary, as a general rule, U.S. advising and training should help local forces become adept at joint, combined, and interagency operations.

U.S. operational aviation advisors can make important contributions even when the host nation lacks an air force.[26] First, operational aviation advisors can help the host nation understand the contribution of air forces to joint, combined, and interagency operations and help determine the kinds of air capabilities (including unmanned systems) that are appropriate for their needs. This is not about building smaller versions of the U.S. Air Force but rather identifying the capabilities that are the best match for the host nation's security needs. In many cases, that will mean procuring simple turboprop or subsonic jet aircraft, helicopters, and UAVs rather than advanced jet fighters.[27] Second, U.S. aviation advisors can help the nation's ground force and joint commanders understand how U.S. or other friendly air forces might team with their ground forces to defeat internal or other threats.

---

[25] For a discussion of recent U.S. Army Special Forces activities, see Linda Robinson, *Masters of Chaos: The Secret History of SF*, New York: Public Affairs Books, 2004.

[26] We use the expression *operational aviation advising* to refer to the collective training and advising U.S. aviators do abroad. Although some airmen prefer the phrase *combat aviation advising*, we use this somewhat broader phrase to capture the full spectrum of their activities. We agree that combat advising is an essential part of their mission, but we do not want to exclude noncombat activities. Indeed, in the current security environment, U.S. advisors (whether U.S. Army Special Forces, U.S. Navy SEALs, U.S. Marines, or USAF personnel) must be given the authority to participate in combat missions if they are going to be effective helping partner nations defeat insurgents and terrorists.

[27] The value of relatively simple aircraft in counterinsurgency operations has been demonstrated in conflicts from Rhodesia to El Salvador. See Corum and Johnson (2003); Roy Nesbit and Dudley Cowderoy, *Britain's Rebel Air Force: The War from the Air in Rhodesia, 1965–1980*, London, UK: Grub Street, 1998; and Jose Angel Moroni Bracamonte and David E. Spencer, *Strategy and Tactics of the Salvadoran FMLN Guerrillas: Last Battle of the Cold War, Blueprint for Future Conflicts*, Westport, Conn.: Praeger Publishers, 1995.

At a higher level, USAF advisors can also help the nation's air staff and ministry of defense develop the organizations, processes, and electronic connectivity necessary to exploit the advantages that air power brings. Accomplishing these more-ambitious objectives will, however, require training more USAF personnel to be operational aviation advisors and international affairs specialists,[28] expanding USAF advising and assistance to the operational and strategic levels, and interacting more frequently and for longer periods with the host nation's forces. An expanded program to develop officers with the appropriate language and cultural skills will be necessary.

For littoral nations, naval advising, training, and equipping will be important as well. Naval forces have been active since 9/11 interdicting the movement of terrorists and weapons, but there are simply too many target vessels for the United States and other major navies to monitor. Local and regional coast guards are needed to monitor and board the thousands of small fishing and trading vessels. U.S. naval advisors are key to training these local navies and coast guards in interdiction, boarding, and related operations. Local forces using small patrol boats can be highly effective in these missions.

The potential demand for land, naval, and air advisors could easily outstrip the supply of SOF personnel who specialize in training and advising. Although some expansion of the SOF advisors is desirable, it will be difficult to greatly increase their numbers. Rather, conventional forces—which already are involved in training foreign partners—will need to take on a greater role. Great caution is warranted here, given the sensitive nature and difficult circumstances associated with many advisory missions. This is particularly true for counterinsurgency, in which the integration of military and civil operations is tricky and essential. That said, conventional forces—carefully trained and prepared for the mission—have the potential to supplement SOF effectively for some training missions.

---

[28] The USAF is expanding its program to develop officers with the necessary language and cultural skills. In this new program, foreign area officers will be replaced by international affairs specialists who will serve as either regional affairs strategists or political-military affairs strategists. See John L. Conway III, "The View from the Tower of Babel: Air Force Language Posture for Global Engagement," *Air & Space Power Journal,* Summer 2005.

Some Caveats on Early Intervention

The preceding discussion focused on the military dimension of assistance, but at all stages, military activities must be carefully integrated with law enforcement, intelligence, and development initiatives. In the early stages of an insurgency, the most important aid will often be that from civilian agencies. Programs to help the partner nation make social, political, and economic reforms are essential at all stages of insurgencies and, if implemented early, have the potential to slow or stop the growth of an insurgency.

In some cases, military assistance will not be possible or appropriate. Governments that are unwilling or unable to address legitimate and fundamental grievances may be propped up temporarily by military assistance, but if that aid is used merely to repress dissent, it is unlikely to have a long-term positive effect. In some cases, governments do not want or need U.S. assistance or will place such constraints on it that it cannot be effective. Finally, some governments and militaries have entrenched cultures of violence, corruption, and oppression. U.S. aid would simply be siphoned off by corrupt officials or used to further abuse their citizenry. In these cases, fundamental reforms will be necessary before U.S. military assistance would be appropriate.

Insurgencies rarely occur in healthy countries. Rather, political violence is most common in countries with weak institutions, oppressive security structures, incompetent and corrupt officials, and limited freedoms. Thus, virtually any country that is a candidate for aid will likely be acting in some way contrary to U.S. values. Deciding whether or not to intervene in some way will usually be difficult. In countries where the armed forces are the most powerful institutions, military contacts might open up avenues to influence the key leaders to undertake reform. Yet the United States has found partner-nation governments and militaries notably resistant to pressure to reform and must assess such opportunities with open eyes.[29] For these reasons, military assistance will often be controversial. The U.S. experience providing

---

[29] Schwarz (1991) argues that the United States regularly underestimates the difficulty of changing foreign institutions.

counterinsurgency assistance to the El Salvadoran government is an instructive and somewhat cautionary tale.

The Power and Limitations of Military Assistance: The El Salvador Experience

U.S. involvement in El Salvador in the 1980s illustrates both the power and limitations of military assistance. The story of the El Salvador civil war and U.S. involvement in it is long and complex, told well in the extensive literature on the topic.[30] This section focuses on a few key lessons.

The roots of the conflict went back decades, some would argue centuries. The exclusion of the indigenous peoples from any meaningful role in the Salvadoran political process, the extreme concentration of power and wealth in the hands of the landed aristocracy, weak government institutions, and the violent repression of dissent all set the stage for periodic violent uprisings.[31] As one former U.S. military group commander in El Salvador observed, "if the campesino didn't like it, he had a couple of options: you could emigrate or you could become part of the fertilizer program."[32]

These various uprisings lacked any cohesive ideology until 1932, when the Communist Party of El Salvador (PCS) was formed.[33] The same year, a Communist organizer, Farabundo Marti, organized a campesino revolt. The government rapidly and decisively put down the

---

[30] Military readers should find the following of particular interest: Bacevich et al. (1988); Max Manwaring and Court Prisk, *El Salvador at War: An Oral History*, Washington, D.C.: National Defense University Press, 1988; and Schwarz (1991).

[31] See Tommie Sue Montgomery, *Revolution in El Salvador: From Civil Strife to Civil Peace*, Boulder, Colo.: Westview Press, 1994; Mario Lungo Ucles, *El Salvador in the Eighties: Counterinsurgency and Revolution*, Philadelphia: Temple University Press, 1996; Elizabeth Jean Wood, *Insurgent Collective Action and Civil War in El Salvador*, Cambridge, UK: Cambridge University Press, 2003; Bracamonte and Spencer, 1995; and Manwaring and Prisk, 1988, for more-complete explorations of the roots of the insurgency.

[32] Col John Waghelstein quoted in Manwaring and Prisk (1988, p. 8).

[33] Montgomery (1992, p. 101).

revolt, killing between 2,000 and 10,000 rebels; Marti was captured, tried, and executed. Although Marti became an icon for the left in El Salvador, the PCS failed to take root beyond radical students and priests.[34] The army seized power the same year and retained it for the next 50 years. During these years,

> two political characteristics held constant: the politics of the regime never threatened the socioeconomic foundations of oligarchic power and the military never allowed the political system to become so open that reformist civilians might actually win control of the government.[35]

LeoGrande continues:

> The process of political polarization in El Salvador began to accelerate in 1972 when the Christian Democrats (PDC) led by Napolean Duarte won the presidential election, but were cheated out of victory by the military's fraudulent counting of the ballots . . . the armed forces unleashed a wave of repression against the PDC which drove most of its leaders into exile. Despairing of the prospects for peaceful change, many rank and file Christian Democrats began looking to the radical left as the only viable opposition.[36]

At the same time, the 1969 war with Honduras brought additional problems for El Salvador. The loss of the Central American Common Market and absorption of 100,000 Salvadoran refugees from Honduras worsened the economic situation, leading a group of Salvadoran communists to conclude that the country was ready for a revolution. They broke from the PCS, forming the Popular Liberation Front. Over the course of the next few years, several other revolutionary groups were founded: the Popular Revolutionary Army, the National Resis-

---

[34] Bracamonte and Spencer (1995, p. 2).

[35] William M. LeoGrande, "A Splendid Little War: Drawing the Line in El Salvador," *International Security*, Summer 1981, p. 30.

[36] LeoGrande (1981, p. 30).

tance, the Central American Workers Party, and the Armed Forces of Liberation.[37]

During the 1970s, the government had little success containing the growth of underground organizations. Private death squads aggressively targeted insurgent and leftist organizations and individuals, killing thousands. The 1977 Public Order Law only made things worse by essentially outlawing political activities. By 1979, in the face of growing political chaos, a reform-oriented group of military officers took power, creating a new regime that included moderate civilians as well. The government, however, quickly became paralyzed and was unable to enact reforms. Civilian members of the government left in frustration.[38]

After the 1979 fall of the Somoza regime in neighboring Nicaragua, many eyes turned to El Salvador. Leftists throughout Central America believed that the success in Nicaragua could be duplicated elsewhere in the region. More importantly, the Cubans believed that a Marxist-Leninist rebellion could succeed in El Salvador and were willing to provide training, arms, and other assistance if the various groups would agree to work together as part of a coalition effort. Despite their animosity toward one another, the Popular Liberation Front, National Resistance, Popular Revolutionary Army, Central American Workers Party, and Armed Forces of Liberation agreed to come together as FMLN.[39]

FMLN began large-scale operations with the 1981 "final offensive," which sought to combine attacks on military garrisons throughout the country with a national strike and rebellion within the armed forces. Although the offensive failed to accomplish these ambitious goals, it "helped to equip the guerrillas to wage a long-term conflict under adverse conditions against a much larger armed forces establishment."[40] Over the next two years, FMLN expanded its influence in the

---

[37] Bracamonte and Spencer (1995, p. 2).

[38] LeoGrande (1981, pp. 30–31).

[39] See Bracamonte and Spencer (1995, Chapters 2–6) for a discussion of the origins of FMLN.

[40] Ucles (1996, p. 19).

eastern, northern, and central regions of the country and humiliated El Salvador's army in many battles. Indeed, as early as 1981,

> the shortcomings of the Salvadoran forces were so pronounced that a [sic] FMLN victory appeared likely . . . the FMLN held the initiative and operated freely in many parts of the country, especially at night.[41]

In the face of these challenges, the United States embarked on a massive military and economic assistance effort to El Salvador, totaling over $1 billion in military assistance alone between 1980 and 1990. Between 1980 and 1983, U.S. aid allowed the El Salvador Armed Forces (ESAF) to grow from 17,000 to 37,000 personnel and increase the number of maneuver battalions from 16 to 43.[42] The Salvadoran Air Force (FAS) received four O-2A Skymaster aircraft (to be used for reconnaissance), six A-37B Dragonfly aircraft for strike, and two C-123 Provider transports during the first half of 1982 alone.[43] Over the next few years, the United States provided the FAS with a substantial helicopter force: 88 UH-1H Iroquois transports, 23 UH-1M gunships, and 14 Hughes 500 armed reconnaissance helicopters.[44] In 1984 and 1985 the United States provided two AC-47 gunships, perhaps the most lethal source of airborne firepower during the war.[45]

How effective were these efforts? It is not an exaggeration to say that U.S. training, equipping, and advising transformed the ESAF. In a short period, the El Salvadoran Army went from a garrison force with little combat power to an army capable of conducting effective battalion-sized offensive operations supported by substantial firepower. The ESAF performed reasonably well during the conventional phase of the

---

[41] U.S. Government Accountability Office, *El Salvador: Military Assistance Has Helped Counter but Not Overcome the Insurgency*, Washington, D.C., April 1991, p. 13.

[42] U.S. GAO (1991, p. 9) and Bacevich et al. (1988, Table 3, p. 5).

[43] Corum and Johnson (2003, p. 333).

[44] Aeroflight, "El Salvador Air Force Unit History: Grupo de Operaciónes Especiales," *World Air Forces* Web site, May 1, 2003a, and Aeroflight, El Salvador Air Force Unit History: Escuadrón de Helicópteros," *World Air Forces* Web site, May 24, 2003b.

[45] Corum and Johnson (2003, pp. 337, 347–348).

war (1980–1984), forcing FMLN to abandon larger-scale operations and return to guerrilla tactics.[46]

ESAF performance, however, declined once FMLN dispersed. The Army never fully embraced counterinsurgency tactics and remained obsessed with large conventional operations, movement by truck or helicopter, and excessive use of firepower throughout the remainder of the war.[47] The emphasis on firepower led Salvadoran infantry units to carry 90-mm recoilless rifles, 81-mm mortars, and 0.50-caliber machine guns on many missions for which assault rifles and medium machine guns would have been sufficient. Carrying such heavy weapons and ammunition in rugged terrain meant that "when contact is made, they don't have the foot mobility to maneuver around and cover the ground fast enough to cut the guerrillas off. They are simply far less mobile than the guerrillas."[48] This reflected a more-systemic problem with equipment. The United States did an outstanding job equipping the ESAF with everything from boots to rifles, but U.S. security assistance "failed to wean the Salvadorians from their conventional mindset," allowing them to purchase heavy weapons "of little utility in a counterinsurgency." For example, "rather than risk disaffecting ESAF by opposing requests for inappropriate hardware, American officers at times succumb to the temptation to go along."[49] Bacevich cites the example of 105-mm howitzers. ESAF used these mainly in harassment and interdiction missions, which the United States had learned in Vietnam are "at best wasteful and at worst counterproductive."[50] Some elite units, such as the Atlacatl Battalion, developed into aggressive and effec-

---

[46] The Battle of Campana Hill on January 4–8, 1985, was one of the last large-scale conventional battles. See Bracamonte and Spencer (1995, pp. 59–66) for a detailed description of the battle and FMLN's shift back to guerrilla tactics.

[47] For assessments of U.S. training and education efforts, see Michael Childress, *The Effectiveness of U.S. Training Efforts in Internal Defense and Development: The Cases of El Salvador and Honduras*, Santa Monica, Calif.: RAND Corporation, MG-250-USDP, 1995; Bacevich et al. (1988); Schwarz (1991); and U.S. GAO (1991).

[48] Colonel Lyman C. Duryea, U.S. Defense Attaché in El Salvador, 1983–1985, quoted in Manwaring and Prisk (1988, p. 318).

[49] Bacevich et al. (1988, pp. 29–30).

[50] Bacevich et al. (1988, p. 30).

tive counterinsurgent forces, but they were the minority.[51] The army at large continued to avoid night operations, and its offensive operations routinely came up empty. American trainers became hugely frustrated, describing ESAF operations as "search and avoid patrols" and deriding ESAF night field-craft as like "a boy scout jamboree—campfires and transistor radios."[52] Although efforts to reform the officer corps and create a professional NCO corps were strongly resisted, U.S. pressure did reduce military involvement in death squads and politics.

FAS developed the ability to conduct company-size air assaults and routine medical evacuations with its rotary wing assets, conduct reconnaissance, and deliver lethal firepower. Using Hughes 500 helicopters and O-2 fixed-wing aircraft as scouts, the FAS would detect FMLN units and attack them with A-37s, AC-47s, and UH-1M gunships. UH-1H helicopters would move ground forces into place under this air cover. But again, these tactics, which proved devastating during the conventional phase, were less relevant once FMLN stopped operating in large formations.[53]

Like the Salvadoran Army, the FAS also had significant limitations. Shortages in trained mechanics and pilots along with an institutional culture that did not value maintenance meant that the FAS had a low percentage of aircraft available for operations. With fewer pilots than aircraft, a typical pilot had to fly three to four aircraft types, preventing the development of deep expertise in one platform. Lack of instructor pilots limited training and led to high accident rates. In the most comprehensive study of the FAS, Corum and Johnson argue that

---

[51] Unfortunately, the Atlacatl Battalion's record of tactical effectiveness was seriously sullied by its horrific human rights abuses. Immediately following one round of training by a U.S. Army Special Forces MTT, the battalion killed six Jesuit priests (on November 16, 1989). The battalion was also responsible for the massacre of 700 villagers at El Mozote in 1981, the killing of "dozens" of villagers in Tenancingo and Copapayo in 1983, and the killing of 68 civilians in Los Llanitos and another 50 at Gualsinga River in 1984. See Montgomery (1994, p. 4); Williams and Walter (1997, p. 143); and Mark Danner, *The Massacre at El Mozote*, New York: Vintage Books, 1994.

[52] Bacevich et al. (1988, p. 37).

[53] Interview with Professor Max Manwaring at the U.S. Army War College, Carlisle, Pa., April 26, 2005.

the UH-1H helicopters made the greatest contribution, followed by O-2 reconnaissance aircraft and AC-47s. FAS pilots lacked the skill to successfully employ the A-37.[54]

El Salvador: Lessons for Future Counterinsurgency Operations

The U.S. experience in El Salvador offers several lessons for future counterinsurgency operations. First, it demonstrated that military assistance is a powerful tool that can prevent the defeat of a partner government facing a serious threat without the introduction of U.S. combat forces.

Second, congressional efforts to limit the U.S. footprint—although frustrating to military participants at the time—proved to be a good thing in retrospect.[55] The limits did prevent U.S. personnel from doing all that they might have in country, and the cost of sending Salvadoran personnel to the United States and Honduras for training reduced the number of soldiers who could be trained. The limits, however, accomplished something hugely beneficial. They prevented the war from becoming Americanized, ensuring that it remained for the Salvadorans to win or lose.

Third, the Salvadoran experience demonstrated once again that air power—both rotary and fixed wing—has much to contribute to rural counterinsurgency operations.[56] Airborne firepower prevented

---

[54] Corum and Johnson (2003, pp. 347–348). Bracamonte and Spencer (1995) give the A-37 higher marks; see pages 59–66.

[55] Manwaring interview, April 26, 2005.

[56] Air power also has much to contribute to urban counterinsurgency, but the emphasis tends to be more on surveillance and infrequent, but vital, precision strike. In Iraq, tactical movement by helicopter is relatively rare within urban areas, although helicopters continue to play vital roles during major offensive operations, for movement between bases, for reconnaissance, and for fire support. Fixed-wing aircraft in Iraq have made major contributions as surveillance platforms; air transport has reduced the volume of materiel moved by road (and thereby reduced the number of personnel exposed to improvised explosive devices); and strike aircraft and armed UAVs have provided on-call fires. For a treatment of air operations in urban environments, see Alan Vick et al. (2000).

FMLN from achieving battlefield victories, helicopter medical evacuation proved a huge morale booster for ESAF soldiers, and airborne reconnaissance platforms were vital sources of intelligence.[57]

Finally, the Salvadoran experience shows the limitations of military assistance. Although the aid prevented an FMLN victory, ESAF refused to reform itself into a professional military optimized for counterinsurgency operations.[58] Thus, it was unable to defeat FMLN on the battlefield or significantly undermine the group's popular support. More broadly, the failure of the Salvadoran government to implement a national-level plan that addressed fundamental social, political, and economic problems meant that the counterinsurgency effort was never fully unified—a critical lapse, as discussed in Chapter Three.

U.S. military aid did enable the ESAF to reach a stalemate with FMLN, lasting roughly from 1984 until 1989. When the Cold War ended in 1989, U.S. and Soviet interest in the conflict waned, and Soviet support for Cuba (and Cuban support for FMLN) largely ended. Ultimately, the conflict was resolved through U.N.-sponsored negotiations.

In the next chapter, we turn our attention to the role of air power in countering insurgencies and, more specifically, how best to prepare USAF for future counterinsurgency challenges.

---

57 Interview with Professor Caesar Sereseres at the University of California, Irvine, May 9, 2005.

58 See Schwarz (1991) for a thorough discussion of U.S. efforts to reform the ESAF.

CHAPTER SIX

The USAF Role in Countering Insurgencies

This chapter considers USAF's role in counterinsurgency operations. Beginning with a brief discussion of the contributions of air power to counterinsurgency, it then moves to consider specific USAF contributions in training, advising, and assisting partner-nation air forces. After briefly describing the 6th Special Operations Squadron (6 SOS) (the sole USAF organization assigned the aviation advising mission), we offer a methodology for estimating global post-9/11 demand for USAF aviation advising capabilities and then move from this to estimate the manpower required for supporting expanded advising activities. The chapter concludes with a set of proposals for enhancing USAF's capabilities for countering insurgencies.

The Role of Air Power in Counterinsurgency Operations

Downplayed, taken for granted, or simply ignored, air power is usually the last thing that most military professionals think of when the topic of counterinsurgency is raised. Until recently, this was as true of airmen as it was for the other services. Writing in 1998, one air power scholar observed that, "[t]o a large extent, the Air Force has ignored insurgency as much as possible, preferring to think of it as little more than a small version of conventional war."[1] Yet, "since at least 1915 . . . the United States has used air power in more than a dozen conflicts against

---

[1] Dennis M. Drew, "U.S. Airpower Theory and the Insurgent Challenge: A Short Journey to Confusion," *Journal of Military History*, October 1998, p. 809.

guerrillas, so-called bandits, and other irregulars."[2] Indeed, nations as diverse as the former Soviet Union and El Salvador have proved that air power is not just useful but essential to counterinsurgency operations.[3] The settings have spanned the globe and include Nicaragua; Greece; the Philippines; Malaya; Southeast Asia; Oman; Rhodesia; El Salvador; Colombia; and more recently, Afghanistan and Iraq. During these operations, air power consistently provided mobility, reconnaissance, and strike capabilities that greatly enhanced the effectiveness of counterinsurgency ground forces.

So why, given this long and accomplished history, is the contribution of air power in counterinsurgency undervalued? There are several reasons.

First, because insurgent movements lack large industrial, transportation, communications, or military centers, they are largely invulnerable to classic air campaigns. It is true that, in some cases, insurgents have been supported by outside powers that possess these traditional attributes. As discussed in Chapter Four, these states are theoretically vulnerable to a more-traditional application of air power (or coercive threats to use it), but this has rarely been done.[4] Rather, in most cases

---

[2] Corum and Johnson (2003, p. 3).

[3] Readers interested in the history of air power in counterinsurgency are referred to Corum and Johnson (2003). Their book is the only comprehensive historical and analytical treatment of air power in small wars. See also Bruce Hoffman, *British Air Power in Peripheral Conflict, 1919–1976,* Santa Monica, Calif.: RAND Corporation, R-3749-AF, 1989, and Mark A. Lorell, *Airpower in Peripheral Conflict: The French Experience in Africa*, Santa Monica, Calif.: RAND Corporation, R-3660-AF, 1991. A unique reference is the six-volume RAND report from a 1963 symposium at which counterinsurgency practitioners discussed the tactical and operational lessons they had learned from the use of air power in unconventional or counterinsurgency operations in World War II, the Philippines, Malaya, and Algeria. See A. G. Peterson, G. C. Reinhardt, and E. E. Conger, eds., *Symposium on the Role of Airpower in Counterinsurgency and Unconventional Warfare*, 6 vols., 1963 (full citations in bibliography).

[4] One of the few examples is from Rhodesia. The neighboring states of Zambia and Mozambique provided sanctuary and support to insurgents seeking to overthrow the white regime in Salisbury. The Rhodesian Air Force and Special Air Service attacked rebel bases and economic targets in these countries. Power lines, ferries, roads, and bridges were all destroyed to disrupt the economies of Zambia and Mozambique and "convince these countries that it was no longer in their interests to play host to the guerillas" (Nesbit and Cowderoy, 1998,

the counterinsurgent government lacked the air capabilities necessary to credibly threaten such action or was itself deterred by the escalatory risks associated with such a move. More typically in the post–Cold War era, insurgents receive support from diasporas, other nonstate actors, or states that are geographically removed or sufficiently covert about their support to avoid direct confrontation with the government fighting the insurgency.[5]

Second, because insurgencies do not present opportunities for an overwhelming application of the air instrument, air power has been used in a less-visible supporting role. Flying intelligence, surveillance, and reconnaissance missions; airlifting troops; evacuating the wounded; and providing fire support for engaged ground forces tend to be taken for granted or undervalued outside aviation communities.[6] Finally, U.S. observers tend to view helicopters, which are used for mobility, reconnaissance, and fire support, as army platforms whether or not they belong to that nation's air force or army. As we argue in an earlier report,[7] it is more accurate and helpful to think of air power—from whatever service—as a partner with ground and other military forces than to emphasize who is supporting or being supported. This is particularly true in counterinsurgency. As discussed in earlier chapters, successful counterinsurgency requires unity of effort across multiple government agencies, including those providing political, economic, law enforcement, and intelligence assistance, as well as the military. Indeed, the only subordination that really matters in counterinsurgency is that all the various activities contribute to the fundamental political strategy. All military operations, including ground, naval,

p. 102). Nesbit argues that, because of these attacks, the leaders of Zambia and Mozambique pressured Rhodesian insurgent leaders to arrive at a peace settlement or risk losing all support. As a result, the rebels were forced "to accept the principle of a general election and to accept the outcome in the world spotlight" (Nesbit and Cowderoy, 1998, p. 110).

[5] A 2001 RAND study identified 44 post–Cold War insurgencies in which state support (usually from a neighboring country) played a major role. See Daniel Byman et al. (2001).

[6] This is particularly true for urban counterinsurgency, in which the role of air power is generally less visible and more constrained.

[7] Pirnie et al. (2005).

and aerospace operations, are therefore best understood as supporting efforts to the overarching political strategy.

How Should We Think About the Role of Air Power in Counterinsurgency?

Although its contributions may be less obvious to the casual observer, the historical record is clear and consistent: Air power has made major contributions to counterinsurgency in a broad range of settings. Interestingly, although many consider air power to be a high-technology instrument that only the richest countries can employ, countries with more-limited resources have used relatively simple systems to great effect. For example, although Rhodesia also flew more-advanced fighter and bomber jets, light civilian aircraft also played an important reconnaissance role during its insurgency. "Pilots and observers, flying low and slow, became adept at spotting guerrilla tracks and signs in the bush and passing the information to the fire force."[8] In El Salvador, a mix of older, relatively simple systems, including AC-47 gunships, OV-2 reconnaissance aircraft, and Hughes 500 and UH-1 helicopters were highly successful in joint air-ground operations. "What always gave the armed forces the edge was air power."[9] At the battle of Campana Hill in 1985, "A-37 attack planes and AC-47 gunships turned the tide of the battle in favor of the besieged army troops, averted disaster, and inflicted heavy casualties on the assaulting guerrilla force."[10]

Neither air forces nor other military forces, however, can by themselves defeat an insurgency, but when used wisely, they can help establish a secure environment within which the other counterinsurgency instruments can work. In this short section, we briefly review some of the major contributions of air power to counterinsurgency, which include limiting the adversary's conventional options, accelerating gov-

---

[8] The fire force was a quick-reaction force that would arrive by parachute from older C-47s and by small Alouette helicopters (Corum and Johnson, 2003, p. 299).

[9] Bracamonte and Spencer (1995, p. 66).

[10] Bracamonte and Spencer (1995, p. 66).

ernment response to insurgent attacks, and seizing the tactical initiative from insurgents. Each of these contributions is essential to military success in counterinsurgency.

Limiting Adversary Conventional Options

Air power constrains the adversary's options from the strategic to the tactical level. Because of its ability to conduct wide-area surveillance and destroy massed forces in the open, air power makes it difficult for insurgents to shift to conventional tactics. It is easiest for air power to do this against mechanized forces in the open, but air power has successfully done this against some light infantry foes in more-rugged terrain and foliage. For example, after several years of FAS pummeling, FMLN abandoned battalion-level operations in 1984 and for most of the remainder of the conflict.[11] Thus, air power can help bound a conflict and deny the enemy some escalation options. Today in Afghanistan and Iraq, airborne surveillance makes it difficult for insurgent forces to move in large numbers or to mass on a target without detection. This allows friendly forces to patrol in small numbers or be stationed in isolated villages without risk of being overwhelmed by a large insurgent force. In short, air power makes it difficult for insurgents to shift to a conventional phase or even to mass for a raid. Air power also limits the options of neighboring countries that might be tempted to intervene in the conflict with conventional forces.

Balancing Insurgent Advantages

Insurgents generally enjoy the advantage of the initiative, choosing the time and place to conduct combat operations. All else being equal, they will choose targets that are isolated, allowing them to attack, then disappear before reinforcements arrive.

A classic example from the Vietnam War was the nighttime attack by the Viet Cong on an isolated village or military outpost. Without air power, government forces had limited means of responding to such attacks. With air power, multiple options open up. The speed and range of aircraft make it possible to respond to emergencies

---

[11] Bracamonte and Spencer (1995).

across distances and terrain that would be slow or impractical to cross on the ground. Ground forces can be moved rapidly by air to reinforce embattled patrols or outposts. Strike aircraft can be placed in orbits over high-threat areas to provide immediate fire support during the most dangerous periods or to be available on ground alert.

Gaining the Initiative

Enduring airborne surveillance of enemy operating areas, when combined with other sources of intelligence, increases the number of opportunities for counterinsurgency forces to take the initiative. For example, such surveillance might lead to the detection of an insurgent base. As a result, a joint force might attack the base, moving to the target by air and providing surveillance, communication, and fire support by air. More routinely, air power has moved patrols deep into enemy terrain, resupplied them, and provided fire support and extraction as required. Air power can help gain the initiative at the tactical level as well. For example, reconnaissance platforms can support patrols by flying ahead to detect potential threats. Once detected, the ground force can maneuver around the threat or call in air or other fires against it. Air power also supports raids by monitoring likely insurgent escape routes and directing U.S. forces to insurgents or weapon caches.[12]

In sum, the unique advantages of air power—its speed, range, persistence, flexibility, and lethality—made it integral to counterinsurgency operations in the 20th century. Large and small nations, using systems ranging from the old and simple to the most advanced, have found it to be essential for countering the inherent advantages of the insurgent. As the United States looks to assist partner nations in countering insurgents, it needs to ensure that their air forces are well trained and equipped for this difficult mission. In the next section, we discuss USAF's role in advising and training these friendly air forces for counterinsurgency.

---

[12] See David Wood, "The 'Poo Hunt': In an Unconventional War, Creative Use of Air Power," *Newhouse News Service*, August 18, 2005; David A. Fulghum, "Combat Aviators Claim 'Non-Kinetic Warfare is Here,'" *Aviation Week and Space Technology*, May 23, 2005a, p. 50; and David A. Fulghum, "USAF F-15E's New Non-Bombing Mission Draws Praise," *Aviation Week and Space Technology*, May 23, 2005b, p. 53.

Current USAF Operational Aviation Advising Activities

The 6 SOS is currently the only organization within the Air Force dedicated to the training and advising of foreign aviation forces. Reactivated in 1994 as a standing advisory force within the Air Force Special Operations Command (AFSOC), this small squadron of approximately 100 personnel trains and advises foreign air forces on the employment of air power for internal defense.

Although the unit initially focused on assisting friendly nations to combat the internal threats of insurgency and drug violence, the role of the 6 SOS has expanded since 2001 to include the training of foreign forces to counter global terrorism. This new focus on counterterrorism has significantly increased the demand for 6 SOS forces and changed both the types and locations of missions that the squadron has been tasked to undertake.

This section presents a brief history of the 6 SOS and an overview of its current mission and organization. It also analyzes some of the trends in operational aviation advisory activity and deployment over the last decade and considers how the demand for more advisors may be linked to future manning and force-structure requirements.

6 SOS History

While the current 6 SOS has only been in operation since 1994, the history of the unit extends back to World War II. In 1944, the 6th fighter group was first established as a small specialized Air Corps combat force trained to support British guerrilla forces operating behind enemy lines in Burma. As part of an integrated air commando group,[13] the 6th fighters provided training and direct air support to the unconventional British forces.[14] The unit was disbanded soon after

---

[13] This group, known as the First Air Commandos, consisted of a composite force of fighters, bombers, transport, glider, and helicopter aircraft, with 523 men and 348 aircraft tailored exclusively to support British Brigadier Orde Wingate and his guerrilla forces (known as Chindits). Michael E. Haas, *Apollo's Warriors: United States Air Force Special Operations during the Cold War*, Maxwell Air Force Base, Ala.: Air University Press, 1997.

[14] Hass (1997) and Capt Timothy Bailey, "Air Commando: A Heritage Wrapped in Secrecy," *Airman*, March 1997.

the end of the war, having established a precedent for the employment of a specialized aviation squadron equipped to train, advise, and assist foreign nations engaged in unconventional warfare.

In 1962, the air commando model was revived when the unit was reconstituted to respond to a different challenge, Soviet-supported insurgencies in the Third World. Focusing at first on training foreign air force personnel in the application of air power in counterinsurgency (in contrast to their original mission of supporting guerrilla forces), squadron personnel (along with the larger 4400 Combat Crew Training Squadron and, later, the Special Air Warfare Center) served as advisors to Vietnamese Air Force personnel, as well as to Latin American; Middle Eastern; African; and, later, Thai and Laotian airmen.[15] As U.S. involvement in the Vietnam War shifted to direct combat with North Vietnamese forces, the role of aviation advisors in training foreign forces was subsumed by the demands of providing close air support for U.S. ground troops.[16] Ultimately, the lines between supporting counterinsurgency and active engagement in counterinsurgency blurred. By 1969, the unit was deactivated, and by the end of Vietnam War, USAF eliminated its entire advisory capability. Following the prevailing attitude of "no more Vietnams," advisory operations were deemed too operationally risky and politically complex,[17] and "counterinsurgency" was eliminated from the DoD lexicon.[18]

Twenty years later, USAF's advisory capacity was reactivated, responding to the requests for assistance from friendly countries, primarily in South and Central America, to help them combat internal threats created from local insurgencies and lawlessness. Prompted by the Goldwater-Nichols Act, which established SOCOM and identified

---

[15] The 4400 Combat Crew Training Squadron was often referred to as "Jungle Jim."

[16] After 1965, the 6 SOS provided air support for ground forces, air cover for transport and interdiction, search and rescue, armed reconnaissance, and forward air control until it was deactivated in 1969. Wray Johnson, "Whither Aviation Foreign Internal Defense?" *Aerospace Power Journal*, Spring 1997, p. 5.

[17] Norman J. Brozenick, *Another Way to Fight: Combat Aviation Advisory Operations,* Maxwell Air Force Base, Ala.: Air University Press, June 2002, p. 43.

[18] W. Johnson (1997, p. 5).

internal defense and development as one of its principal missions, the Air Force created a unit within AFSOC in 1994 that would specifically dedicate itself to foreign internal defense. The unit was designed, following the model of earlier aviation advisory units (in fact, many of the early planners were veterans of the U.S. aviation advisory missions in southeast Asia and Latin America), to be an integrated unit, with a broad base of aviation expertise and regionally oriented, combat-trained personnel specifically trained to assist foreign forces.[19] This unit's structure and its mission, organization, and operations have remained largely the same over the last decade. Only its size has changed since it was reactivated, growing from 47 to 99 between 1994 and 2000, and the types and locations of its advising and training missions have evolved since 9/11.

Mission

The current mission of the 6 SOS is to assess, train, advise, and assist foreign aviation forces in air power employment and sustainment and to integrate these assets into joint, multinational operations. The primary context for this mission continues to be foreign internal defense (the training of host nations to deal with internal threats), yet it now also includes coalition support operations (in which foreign units are engaged in contingency actions or regional warfare) and unconventional warfare (in which foreign aviation forces are trained to support guerrilla operations).

In each of these contexts, the squadron's efforts focus on tactical and operations support. Rather than providing basic flying skills or weapon upgrade training, squadron missions are geared to providing collective training in applied tactics, techniques, and procedures. This training is intended to advance a host nation's tactical aviation skills and to improve the availability, safety, and reliability of its existing capabilities.

---

[19] Like their air commando predecessors, the unit adopted the term "combat aviation advisors" to emphasize its ability to provide combat assistance to foreign nations when necessary.

Training packages are tailored to specific conflicts or tactical situations, with instruction provided in the host country, in the host language, utilizing indigenous aircraft and equipment. Specific types of training include combat search and rescue, fixed- and rotary-wing tactical airlift, medical evacuation, and air attack. This tactical training is most often combined with assistance in aviation support operations, which includes aviation maintenance, supply, munitions, ground safety, life support, personal survival, evasion, resistance and escape airbase defense, medical support, and combat command and control. The 6 SOS often operates with other SOF (e.g., Army Special Forces or Navy SEALs)

The 6 SOS also provides operational advice and assistance to foreign aviation units and to U.S. combatant commanders on force integration. Squadron personnel advise host nations on how to better organize and employ air power and support U.S. commanders in planning and integrating coalition activities.[20]

Typical Engagements

Squadron personnel provide tactical and operational support to host nations through four key tasks, which often occur sequentially: assessment, training, advising, and assisting. A typical 6 SOS engagement begins with an initial assessment mission, during which aviation advisors evaluate the host nation's aviation capabilities and limitations. These assessments may cover aircrew capability and safety, aircraft airworthiness, resource availability, and operational potential. Assessments are then followed by training or exercise missions, enabling the host nation's aviation forces, usually through a "train the trainer" technique, to employ a particular tactic or skill. Later, advising missions are conducted to prepare the host nation to apply these tasks within a particular operational context, either to engage with hostile forces or to integrate its forces into coalition operations. Finally, some engagements may include assistance missions in which U.S. advisors support

---

[20] Christopher Bolkom and Kenneth Katzman, *Military Aviation: Issues and Options for Combating Terrorism and Insurgency*, Washington, D.C.: Congressional Research Service, January 24, 2005, p. 9.

foreign nations by directly participating in a tactical combat operation, contingency, or event. Direct-assistance missions are rare, however, and require the authorization of the President or Secretary of Defense.

Squadron Organization

To carry out these missions, 6 SOS is organized into six flights with two flights each dedicated to CENTCOM and U.S. European Command (EUCOM) and one flight each for U.S. Southern Command (SOUTHCOM) and U.S. Pacific Command (PACOM).[21] Each advisory flight consists of personnel with a variety of aviation expertise, representing 32 Air Force Specialty Codes. Skills include fixed- and rotary-wing tactical flying, aircraft maintenance, command and control, airbase defense, aerospace medicine, and personal survival, allowing the squadron to create individual teams with a broad base of air power support.

When deployed, members of the regional advisory flights form operational teams, called operational aviation detachments (OADs), whose size and composition are tailored to meet specific mission needs. OAD-A teams are formed by pilots, aircrew, maintenance personnel, and special tactics specialists to provide tactical training and advice to host nations. OAD-B teams are created to provide infrastructure support in such areas as command, control, and communications; logistics; and administrative and medical support. Some teams have as few as two members; however, a typical or notional mission has 13 people on its OAD-A team and 10 members on an OAD-B team.

Depending on the manpower the squadron has available, advisory teams may be augmented by personnel from outside the squadron or may cross-attach members from other theater flights to obtain the necessary number of trainers and the required expertise for a particular mission. Two or three augmentees typically supplement each 6 SOS mission.

---

[21] Each flight has approximately 13 assigned personnel.

Manning and Training

Authorized manning for the unit is 109 enlisted officers and civilians. The number of assigned personnel has yet to reach this, however. The unit has averaged between 87 and 99 assigned personnel from 2000 through 2005.

This low rate of assignment may be attributed to the squadron's selective screening process and limited pool of applicants. Qualified applicants are required to have expertise in a particular Air Force Specialty Code (usually as a qualified instructor) and relevant language skills. They must also be able to undergo SOF training, live with host-nation forces in austere environments, and interact with foreign military leaders. These factors often limit potential recruits to older volunteers who have had several years of experience, either in conventional Air Force or other SOF units, and who have both a unique skill set and interest in training foreign forces. Given both the limited visibility of the 6 SOS mission and the lack of a clear career path for officer or enlisted personnel, the number of potential candidates is quite small.

The availability of manpower for missions is further limited by the amount of time that squadron personnel are required to devote to training. Before conducting their first mission, new personnel undergo approximately six months of initial academic instruction and experience-based training.[22] Squadron members are also required to take a five-week integrated skills training course, which provides training in the combat survival techniques, advanced weapon training, and antiterrorism skills (such as defensive driving and specialized weapon training) necessary to operate in remote locations without standard force protection.

In addition to this initial training, all squadron personnel are required to maintain their language competencies through formal classes or on-site interactions with host countries and to sustain their technical expertise to ensure that they remain qualified to provide training in a wide variety of environments. Pilots must have regular

---

[22] This initial training includes instruction in foreign internal defense and combat doctrine, cross-cultural communications and language skills, and advanced technical training on the types of aircraft and equipment used in advisory missions.

flying time in both USAF aircraft and the foreign aircraft that they may operate during deployments to retain their instructor qualifications. Maintenance personnel must similarly obtain certification on foreign aircraft to evaluate the maintenance and support functions host nations provide. Moreover, pilots, maintenance personnel, and the entire aviation advising team are often required to participate in two or more weeks of predeployment training before each individual mission to prepare for the specific environment and type of forces with which they will engage.

Trends in Operational Aviation Activity Since 9/11

Since 2001, the squadron has executed 49 missions in 26 nations. A typical mission deploys ten squadron personnel for 25 days.[23] This operational tempo is much higher than it was during the previous five years, when operational advisory teams completed 37 missions to only 19 countries, with an average of seven personnel deploying for 19 days.

The location of many of these missions has also shifted, from SOUTHCOM and PACOM to CENTCOM and EUCOM, as these combatant commands, particularly in regions of Central Asia and Africa, have become the focus of U.S. counterterrorism operations. In addition, the types of countries visited have changed. Table 6.1 shows the locations for missions between 1996 and 2000.

As Table 6.1 illustrates, most countries receiving operational advisory support in the late 1990s had long-standing political and military relationships with the United States (such as Colombia, Jordan, and Korea). In contrast, Table 6.2 shows that, since 2001, most missions have occurred in countries that have not had a significant U.S. military presence (such as Kyrgyzstan, Yemen, and Niger) or have not received operational aviation advisory assistance since the founding of

---

[23] Length of mission is calculated as the number of days the main 6 SOS team deploys for training, beginning with the day of departure from Hurlburt Field, Florida, to the day the team returns. We did not include spin-up training, predeployment site surveys, advance team visits, or augmentee time in these calculations.

Table 6.1
Locations of 6 SOS Missions, 1996–2000

CENTCOM	PACOM	EUCOM	SOUTHCOM
Eritrea	Korea	Botswana	Colombia
Jordan	Indonesia	Poland	El Salvador
Kenya	Thailand	Rwanda	Ecuador
Kuwait		Tunisia	Peru
			Venezuela
			Bolivia
			Costa Rica
			Paraguay

Table 6.2
Locations of 6 SOS Missions, 2001–2004

CENTCOM	PACOM	EUCOM	SOUTHCOM
Afghanistan	Korea	Azerbaijan	Colombia
Jordan	Philippines	Georgia	Dominican Republic
Kazakhstan	Sri Lanka	Hungary	Ecuador
Kyrgyzstan	Thailand	Morocco	Peru
Pakistan		Niger	Paraguay
Qatar		Poland	
Tajikistan		Romania	
Uzbekistan		Slovenia	
Yemen			

the 6 SOS in 1996 (such as Georgia, Pakistan, and the Philippines). In fact, among the 17 countries the squadron visited in CENTCOM and EUCOM from 2001 to 2004, only two countries were visited during the previous period.

Moreover, the types of missions the squadron has undertaken have changed. Since 2001, there have been more training missions than military exercises or assessments, with a greater emphasis on in-depth training in tactical capabilities, such as combat search and rescue techniques and night-vision goggles. Among the list of countries receiving operational advisory support, seven (one or two key countries from each region) have received an intensified training effort or major mission, which consisted of three to five visits during a four-year period. The majority of other, primarily less-developed, countries receiving advisory support received one or two "minor missions" during this period, which focused on initial air power capability assessments and the establishment of new relationships for potential future missions.

Among these missions have been a number of successes in which foreign forces were able to obtain effective combat search and rescue and/or night-vision goggle capabilities allowing them to rescue friendly forces and conduct independent counterterrorism missions successfully or were able to improve airlift mobility to remote regions. In other cases, missions have succeeded in achieving greater interoperability between U.S. and foreign forces, and still others have opened new areas of access for future contingencies.

Advising and assisting missions are, however, rarely straightforward. In several cases, the internal politics of the host nation, a lack of sufficient resources, or a reluctance to accept U.S. assistance caused training to take longer than expected; required more repeated missions; and in some cases, led to postponement or failure of the mission.

Both trends toward more-intensive training in key countries and an increase in the number of initial visits to less-developed countries have increased the length and size of missions. Higher-level tactical training has required longer and more-frequent visits from 6 SOS personnel to establish and maintain new capabilities than would more broad-based military exercises, and initial assessment visits to developing countries in remote locations have been more labor intensive than assessments or exercises in countries with which the United States has longer relationships. Figure 6.1 illustrates how the size and length of missions have increased since 2000, with length of mission increasing most precipitously, growing from nine people per mission for a total of 13 days to 14 people per mission for 29 days in 2003.

Figure 6.1
Changes in Number and Length of Missions and Number of Personnel, 1996–2004

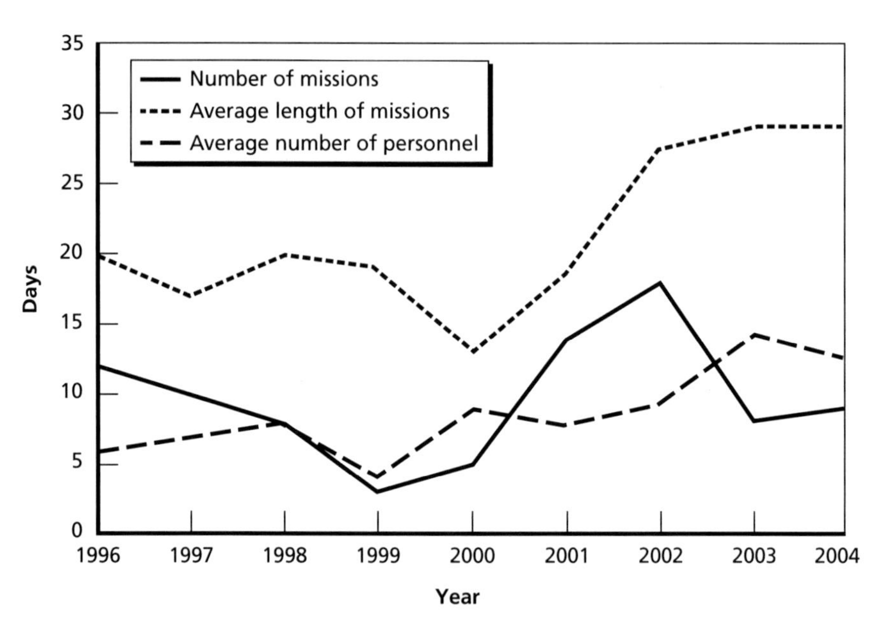

RAND *MG509-6.1*

To represent the combined effects of these trends on the manpower of the squadron more accurately, we calculated the number of "person days" deployed for each mission by multiplying the number of 6 SOS personnel deployed on each mission by the length of the mission.[24] The number of person days per year was then totaled. Figure 6.2 shows that the squadron's deployment levels have more than quadrupled since 2000, growing from 669 to 3,882 person days per year in 2004.[25]

---

[24] The number of personnel includes members of the squadron only. It does not include augmentees from other units who may have been included in the mission. The length of the mission represents the date the main body of the team departs Hurlburt to the day it returns to base.

[25] Person days peaked at 4,814 in 2002, in the aftermath of the 9/11 attacks.

Figure 6.2
6 SOS Number of Person Days Deployed per Year

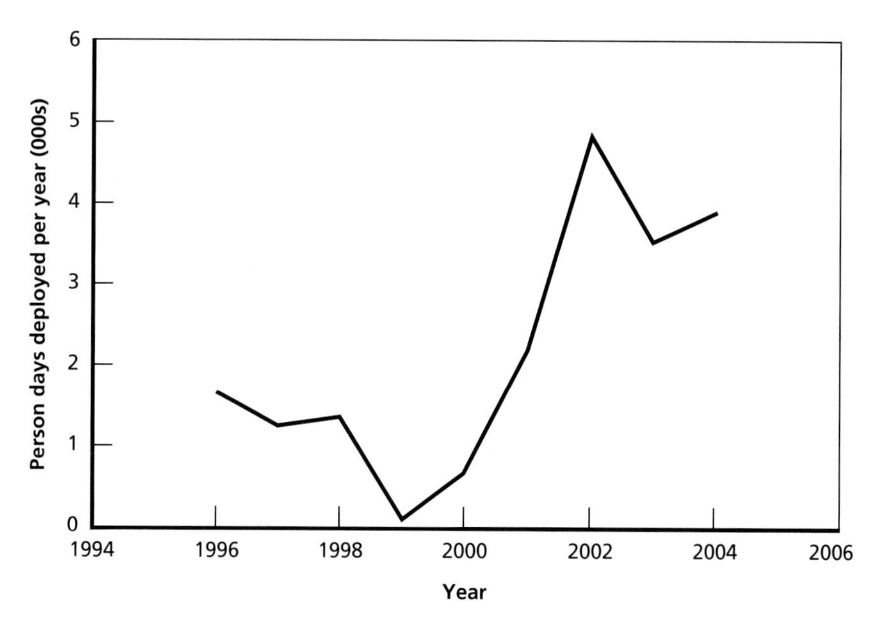

RAND *MG509-6.2*

How this operational tempo has affected the squadron's ability to respond to requests for future operational aviation advisory missions is difficult to quantify. While the 6 SOS does not maintain records of unfilled requests, squadron personnel have estimated that at least 58 percent of the requests for forces that they have received have been turned down, primarily due to a lack of manpower.[26] One former 6 SOS commander has placed this number even higher, indicating that "more requests are received [by the unit] in one quarter than can be met in an entire year."[27] Therefore, current demand may lie somewhere between 200 and 400 percent of capacity. In the next section, we present an approach for estimating the demand associated with the precautionary strategy we argued for earlier in the report.

---

[26] RAND interviews with 6 SOS personnel, March 17–18, 2005.

[27] Brozenick (2002, p. 48).

Estimating Demand for Operational Aviation Advising

Increased emphasis on early involvement in counterinsurgency will require increasing USAF's capabilities for conducting operational aviation advising missions. Estimating this additional demand is analytically challenging. The future is intrinsically uncertain, and projections of future demand must be built on assumptions about the nature of the insurgency challenge and the capacity of U.S. forces. While challenging and fraught with uncertainty, such an analysis is necessary if USAF is to be appropriately prepared for future counterinsurgency operations.

Reduced to essentials, the calculation of future operational aviation advisory demand must incorporate the answers to three questions:

- How many insurgencies will attract U.S. involvement?
- How much USAF operational aviation advisory capacity may be required in each case?
- What is the relationship between USAF's total advising force and the number of missions that can be conducted?

The project team has developed approximations for all three of these to arrive at a parameterized projection of demand for USAF operational aviation advisory capacity.[28]

The number of insurgencies that might attract U.S. and USAF involvement is primarily a function of broader geopolitical and social trends. The total and steady-state levels of USAF involvement are therefore fundamentally unknowable to USAF planners. Lacking data about the future, the next best option is to base planning factors on data from the recent past. The project team used four databases to estimate the number of insurgencies likely to occur in the relevant planning

[28] Time and resource constraints prevented the team from exploring these issues in great detail. USAF leaders will obviously require much more detailed analysis to support policy and programmatic decisions. However, we believe the following analysis effectively captures the basic parameters of the challenge USAF faces.

period.[29] First, we used the World Insurgency and Terrorism database and the Global Terrorism/Insurgency Events databases maintained by Jane's Information Group to compile information on the internal security situation in each nation-state around the world.[30] Jane's updates each database daily to reflect new information. The World Insurgency and Terrorism database tracks the status and activities of virtually all the world's recognized subversive movements.[31] The Global Terrorism/Insurgency Events database tracks activities conducted by terrorist groups or insurgents.[32] To validate and cross check these databases, we also consulted the Uppsala Conflict Database and *Human Development Report 2005* from the United Nations. The Uppsala database summarizes and codes all the world's ongoing conflicts on a number of scores, including yearly fatalities, which helped the team ensure that it properly coded insurgencies as active or latent. The Human Development Report also includes a great deal of data on conflicts and socio-economic indicators for each of the world's states.

By drawing on these four sources of data, the project team was able to develop a reasonably complete and current picture of the number and types of insurgencies around the world. Our analysis of the data suggests that approximately 82 of the world's 191 U.N. member states are beset by some form of active or latent insurgency. Of this number, approximately 35 are beset by insurgencies that are related to militant Islamism—a point of strategic significance, as noted in Chapter Two.[33] Fourteen of these states face active militant Islamist insurgencies, while

---

[29] RAND summer associate Christopher Darnton originated this approach and performed the initial data analysis.

[30] Jane's Information Group, "World Insurgency and Terrorism," Web site, last accessed September 2005; Jane's Terrorism and Insurgency Centre, "Global Terrorism/Insurgency Events," Web site, last accessed September 2005; University of Uppsala, "Uppsala Conflict Database," online database, Uppsala, Sweden, updated annually; and United Nations Development Program, *Human Development Report 2005*, New York, 2005.

[31] Jane's "World Insurgency and Terrorism" site.

[32] Jane's "Global Terrorism/Insurgency Events" site.

[33] See Appendix A for a list of these insurgencies.

another 21 face latent or nascent militant Islamist subversion.[34] Our analysis suggests that approximately 47 states are being afflicted by insurgencies that are not related to militant Islamism; among these, 15 are active, and the remainder (approximately 32) are latent or nascent.

These estimates allowed the project team to make some initial parametric assumptions about the future insurgency challenge. If the insurgency threat remains roughly the same magnitude over the relevant planning period, the data summarized above are a rough indication of the number and types of insurgencies in the steady-state future. However, the insurgency threat might also grow or decline by some percentage or the proportional representations of militant Islamist groups and active versus latent groups or other characteristics might change.

For this analysis, our project team assumed that the magnitude of the insurgent challenge would remain broadly similar to the situation existing today or that, at the very least, USAF planners should assume that insurgencies will be no less prevalent in the future than they are today. This is not to say that USAF should expect to be involved in every future insurgency. U.S. policy toward some insurgencies will be benign, or even supportive; even when the U.S. government sees an interest in the defeat of an emerging insurgency, the political-military context in the afflicted nation may preclude U.S. involvement. In fact, our analysis finds that the United States today has some level of security assistance relationship with approximately 80 percent of the states afflicted with active or latent insurgencies.[35] This suggests that USAF planners could assume, parametrically, that the political-military con-

---

[34] Distinguishing between active and latent insurgencies is, of course, quite difficult. The distinction the project team drew was that when public sources named one or more active groups opposing a government, with an armed strength estimated in the hundreds, we considered an active insurgency to be under way. When there was a simmering conflict but no generally recognized movement per se or only weak groups with little armed strength (i.e., largely terrorists or bandits), we considered this to be a latent insurgency. This distinction is admittedly imperfect, but it is the best the data will support.

[35] The team compared country data from the Foreign Military Training Reports from the departments of State and Defense for FYs 2002–2004 with the insurgency data described above and found some indication of assistance in 67 of the 83 cases.

ditions for providing operational aviation advisory assistance will be conducive in approximately 80 percent of future cases. Thus, having identified 35 insurgencies of interest, USAF planners could reasonably expect that assistance will be appropriate in 28 of those.

After establishing an assumption about the number of insurgencies likely to occur in the relevant planning future, USAF planners must determine how many USAF operational aviation advisory assets might be involved in each case. Here again, it is possible to gain some insight into future demands by examining recent trends. Data from the 2002–2005 Foreign Military Training Reports and after-action reports from the 6 SOS suggest that there are essentially three levels of USAF involvement in operational aviation advisory assistance. The lowest level is approximately one advisory mission per year to the partner nation, typically lasting 30 days and involving 15 USAF personnel. These small missions typically focus on assessment, relationship-building, and assisting the partner military with a very narrow aviation skill or task, such as maintaining a particular type or variant of helicopter.[36]

The second level of involvement comprises, on average, two advisory missions per year to the partner nation, each typically lasting 30 days and involving 15 USAF personnel. This higher level of involvement allows USAF to transfer more-complex aviation skills and tasks to the partner military, such as search and rescue, and to help the partner sustain these capabilities more effectively.

The third level includes two shorter missions and a much larger training mission lasting an average of four months and involving 30 personnel. This higher level of involvement allows USAF to transfer complex aviation capabilities to partner militaries, such as the ability to conduct sizable air assault operations against an insurgent adversary.

In the next section, we present a rough metric for estimating the manpower requirements associated with various levels of involvement.

---

[36] The 6 SOS and other advisory organizations have, through experience, determined how much effort is necessary to give the partner military a specific tactical capability. The more-fundamental question of how much advising and training is necessary to achieve operational- and strategic-level effects against insurgents is much tougher to answer. We know of no rigorous analytical method for addressing the latter question.

Estimating the Personnel Required for Aviation Advising Missions

The 6 SOS experience provides a basis for estimating personnel requirements for various aviation advising options. Using after-action reports and other data from 6 SOS, we developed a metric that relates inputs (personnel assigned) and outputs (mission days deployed per year).[37] To calculate this, we used deployment data for the unit from 2002, 2003, and 2004. We divided the average mission days by a theoretical maximum for deployment. In this case, we used the average number of personnel assigned to the unit for these years multiplied by 180.[38] The equation is as follows:

$$\frac{\underset{\text{(average 2002–2004)}}{\text{Mission days}}}{\underset{\text{(average 2002–2004)}}{\text{Personnel assigned}} \times \underset{\text{(180 days TDY)}}{\text{Goal}}} = \text{Deployment efficiency ratio.}$$

Using the 6 SOS data, we get

$$\frac{4{,}001}{96 \times 180} = 0.23.$$

Rearranging the equation allows us to estimate the size unit required to meet a particular operational tempo goal[39]:

$$\frac{\text{Mission days per year goal}}{41.4} = \text{Manpower required.}$$

[37] We assumed that the 6 SOS use of manpower is representative, neither exceptionally efficient nor inefficient. Determining real unit manpower requirements will require a detailed analysis by manpower and personnel specialists. What we offer here is simply a rough heuristic to give senior USAF officers some sense of the relationship between the demand and the personnel required to support that demand.

[38] USAF policy strives to keep individual deployments under 180 days per year, so we used that number as the theoretical maximum in our calculation.

[39] For an explanation of how the final equation is derived, see Appendix C.

Applying the Metric: Manpower Requirements for an Illustrative Precautionary Strategy

As discussed above, USAF has low-, medium-, and high-level options for assistance. A counterinsurgency strategy that sought to expand USAF advisory assistance would likely go through multiple phases, beginning with widespread low-level activities, then adding deeper involvement in the countries having the greatest need and/or opportunity. Below is one possible assistance profile:

- Phase One
  - 28 countries each receive one minor mission (15 personnel, 30 days) per year
- Phase Two
  - 18 countries each receive one minor mission per year
  - ten countries each receive two minor missions per year
- Phase Three
  - 12 countries each receive one minor mission per year
  - 12 countries each receive two minor missions per year
  - four countries each receive one major mission (31 personnel, 120 days) and two minor missions.

Using the metric described earlier, we can calculate mission days deployed in each phase and use that to estimate manpower requirements. For example, Phase One would deploy 15 personnel for 30 days to 28 countries: 15 × 30 × 28 = 12,600 mission days. Using the metric described above to determine the manpower required, we find that

$$\frac{12{,}600}{41.4} = 304 \text{ personnel.}$$

Table 6.3 presents the mission days and personnel associated with each of the phases for our strategy.

Table 6.3
Manpower Required to Meet Various Mission-Day Goals

Phase	Mission Days per Year	Manpower Required
One	12,600	304
Two	17,100	413
Three	34,680	838

This example is purely illustrative.[40] The number of opportunities for assistance might be much smaller or larger, and the mix of minor and major missions could vary greatly from what is shown here. Clearly, a strategy along these lines is quite ambitious, making great demands on USAF manpower. The first phase alone would require more than tripling the number of operational aviation advisors available for deployment.

Expanding and Deepening USAF Capabilities to Counter Insurgencies

The USAF possesses the world's most advanced aerospace combat and support capabilities, superbly trained officers and NCOs, and support institutions that are the envy of the world's air forces. These capabilities are, however, individually and collectively focused on the service's primary mission of conventional warfighting. As discussed earlier in this chapter, USAF also has outstanding operational aviation advising capabilities and counterinsurgency expertise resident within 6 SOS. AFSOC and its subordinate units (special tactics and AC-130,

---

[40] A unit's ability to generate mission days is a function of a variety of factors, including its administrative support, planning requirements for each mission, and other training and TDY demands. A unit with a less-efficient deployment ratio of 0.18 instead of 0.23 would require more personnel to accomplish the same mission. This unit would require 386 personnel for phase one. In contrast, a unit with a more-efficient deployment ratio of 0.28 would require only 248 personnel for phase one.

MC-130, MH-53, and MH-60 squadrons) also possess capabilities well suited for counterinsurgency. Perhaps most important, AFSOC personnel have experience operating in a wide range of circumstances that present challenges that are similar to those of insurgent situations, including noncombatant evacuations, unconventional warfare, SOF direct action, and combat support of conventional operations.

The future effectiveness of USAF in counterinsurgency is primarily a function of how well these broader institutional capabilities, both within AFSOC and in the wider Air Force, can be refocused and recalibrated for counterinsurgency—and particularly how they can be used to support the air arms of friendly nations. In this section, we propose several steps that USAF could take to enhance its contribution to U.S. counterinsurgency efforts.

Make Counterinsurgency an Institutional Priority

USAF counterinsurgency capabilities will neither deepen nor expand without the strong support of the senior leaders. Only they—in speeches, policy guidance, and programmatic decisions—can convince the institutional USAF that counterinsurgency matters and that it will receive the attention and resources it deserves. Major speeches and new counterinsurgency vision and policy statements will be essential to communicate that the senior leaders are indeed committed to this new emphasis. In particular, only a major push from the senior leaders can overcome the perception that this functional and career area is and will remain a low priority, unlikely to lead to promotions and other opportunities. Speeches and policy statements, of course, must be followed up with significant concrete steps: changes in personnel assignments and promotions, creation of new organizations, expanded funding, and the like.

Create Organizations and Processes to Oversee USAF Counterinsurgency Efforts

If counterinsurgency is going to become a major mission for USAF, it will need to be represented in institutional policy, planning, budgeting, education, and doctrine development processes and in its organizations. In the past, counterinsurgency issues were typically relegated to SOF divisions and branches on the Air Staff and at major commands. Although SOF personnel were generally more conversant with counterinsurgency than their conventional counterparts, relatively few have had much specific counterinsurgency education or experience. SOF officers were generally more conversant because of their interaction with Army SOF and from AFSOC's historical association with counterinsurgency. It is difficult to raise counterinsurgency issues and concerns, unless specifically in support of another service, to senior USAF leaders because USAF does not have counterinsurgency doctrine or forces. To ensure that the appropriate expertise is available to commanders, counterinsurgency desks, branches, or divisions will need to be created at major commands. To develop policy and ensure that USAF-wide efforts to enhance counterinsurgency capabilities remain visible to the senior leadership, it may be necessary to create an air staff division that is responsible for counterinsurgency. This division would ensure that USAF counterinsurgency programs are integrated to support the counterinsurgency vision established by the senior leadership, in the same way that the global Concepts of Operation (CONOPS) Champions integrate programs to support such CONOPS as Global Strike. The Air Force Secretariat, which has responsibility for Foreign Military Sales and related programs, will also need counterinsurgency expertise on its staff.

Although these changes are necessary, they will not be sufficient to ensure that counterinsurgency receives the attention and resources necessary over the long term unless a counterinsurgency educational foundation is also established from commissioning sources to the highest levels of professional military education. AFSOC's commanding general is likely to be an effective advocate for a counterinsurgency advisory wing in AFSOC but is not well placed to advance counterinsurgency capabilities throughout the broader USAF, especially in an era of downsizing and transition to the warfighting headquarters concept. A center for counterinsurgency or, perhaps more broadly, for irregular conflict (counterinsurgency, counterterrorism, irregular warfare, etc.) could be created to coordinate counterinsurgency capabilities across USAF. Such a center, headed by a two- or three-star general, would take the lead on developing concepts and doctrine for irregular conflict; be the proponent for related programs; and raise equipment, force structure, and personnel issues to the senior leadership. Ideally, it would have liaisons from the other services and relevant agencies and would serve as the USAF focal point to the broader counterinsurgency community. This center would probably need to be a major division of the U.S. Air Force Warfare Center under Air Combat Command if it were intended to influence capabilities across the mainstream USAF. We realize this runs counter to the historic practice of placing such organizations under AFSOC, but AFSOC does not, and is not likely to, have sufficient capacity to handle all irregular conflict demands.

Develop and Nurture Counterinsurgency Expertise Throughout USAF

Although there is significant counterinsurgency expertise in AFSOC and in pockets throughout USAF, the number of true experts is well short of what would be needed to meet these new demands. Operations in Iraq are a mixed blessing on this score. To be sure, thousands of USAF personnel now have combat experience in that country and firsthand knowledge of insurgency. On the other hand, it is not clear how much of the Iraq experience is applicable to counterinsurgency more broadly. The United States entered that war with little counterinsurgency expertise in the military or other agencies and with no integrated DoD, let alone interagency, "doctrine," concepts, or processes for counterinsurgency. As a result, many aspects of U.S. operations were inconsistent with lessons learned from earlier conflicts, and the United States has yet to gain traction on the problem. There is much to learn from the Iraq experience—particularly with respect to insurgent strategies and tactics that are already being exported to Afghanistan and are likely to appear elsewhere. Whether U.S. strategy and tactics in Iraq will be useful in other settings remains to be seen.

To ensure a grounded and balanced understanding of insurgency, USAF will need to take steps to expand opportunities for formal education on the social, psychological, cultural, political, security, and economic aspects of insurgency. It is particularly important to study the history of insurgency across the globe to be able to distinguish between characteristics that are idiosyncratic to a particular conflict and those that are universal. Substantial classwork on insurgency should be required for the Air Force Reserve Officer Training Corps; the Air Force Academy; and all phases of Air Force professional military education, from Squadron Officer School to Air War College. At the least, this should cover the best analytical and historical treatments of the topic with an eye toward conveying an understanding of insurgency as a deeply complex sociopolitical phenomenon, as well as the fundamental principles of counterinsurgency. The Chief of Staff's professional reading list should include at least one of the counterinsurgency classics. The USAF Special Operations School curriculum addresses asymmetric warfare and could be easily expanded to offer more in-depth treatment of insurgency. In 1990, Steve Hosmer proposed creating a joint counterinsurgency institute that would offer a 10- to 12-week basic course, as well as more-advanced and -specialized classes. Sixteen years later, that idea remains timely and good.[41] USAF officers working on graduate degrees in the social sciences and history should be encouraged to write theses on insurgency, and USAF professional journals should encourage article submissions on this topic. It may also be helpful to establish a center to develop counterinsurgency concepts, doctrine, and tactics. During the 1980s, USAF and the Army created the Center for Low Intensity Conflict to do this. A joint and interagency center along these lines could offer another avenue for nurturing counterinsurgency expertise in USAF.[42]

Create a Wing-Level Organization for Aviation Advising

Education, of course, is necessary but hardly sufficient to produce effective counterinsurgency practitioners. USAF personnel also will need experience with real-world insurgencies. This experience may be developed while working on major command staffs or serving in embassies and through other means, but, in our judgment, the single most effective means of expanding this expertise is through the creation of a wing-sized organization dedicated to aviation advising. Such an organization

[41] Hosmer (1990, pp. 21–22).

[42] We thank RAND colleague Bruce Hoffman for this suggestion.

will be necessary to implement a precautionary strategy and would be the core of an expanded counterinsurgency capability in USAF.

The Air Force does not have the force structure and personnel to support a precautionary strategy as outlined in this report. USAF's cadre of counterinsurgency specialists is the fulcrum on which the service's broader capabilities are pivoted to support friendly air arms. This cadre is tiny, both in absolute terms and when compared with the counterinsurgency challenge the nation and USAF face.

USAF can and should move quickly to remedy this situation. In particular, it should expand its aviation advisory capacity to at least wing strength.[43] The focus of this unit would be to help partner air forces prepare for and conduct aviation counterinsurgency. The most logical place for this new wing would be in AFSOC, which is the only element of USAF formally tasked with the counterinsurgency mission.[44]

The establishment of an aviation advisory wing structure would add both depth and breadth to USAF's existing capacity, allowing the service to conduct operational aviation advisory missions to a greater number of partners and allies, more often, for longer periods, and addressing a broader spectrum of operational challenges. It would also allow aviation advising to expand beyond activities focused on flying, maintenance, communications, force protection, and associated air base–level operations to include institutional aviation advising. By this we mean helping a partner air force prepare for counterinsurgency by advising its air staff and ministry of defense on the capabilities they need to acquire; on concepts and strategies for integrating air power into joint and interagency counterinsurgency operations; and

---

[43] We are not the first to make this recommendation. Col Norman Brozenick proposed a Combat Aviation Advisory Group in his 2002 monograph.

[44] The project team identified three options for naming the proposed wing. First, in accordance with current USAF and AFSOC policy, it could be given a generic numbered special operations wing designation. Second, a more functionally descriptive designation would be a numbered aviation advisory wing. Third, and most attractive to many in the community, USAF could name the new wing after its historical predecessor, the 1st Air Commando Wing.

on broader institutional processes and issues, such as resource management, manpower, personnel, and training.

Greater Depth. The core of a wing organization would be a number of aviation advisory squadrons, organized similarly to the current 6 SOS.[45] The prospective wing would oversee four to eight such squadrons.[46] The squadrons would be oriented on specific regions of the world to support long-term cultural and language familiarity; knowledge of counterinsurgency and air power challenges in the region; and working relationships with USAF regional component staffs, combatant command staffs, and U.S. embassy country teams in the region. Ideally, some squadrons would be based in their assigned regions or at least have forward-deployed elements in theater. The squadrons would serve as repositories of regional information and experience, provide administrative and institutional support to their subordinate teams, and ensure that missions are fully coordinated with the broader political and military strategies of the host nation and the U.S. government.

The squadron OAD-A teams would execute individual operational aviation advisory missions, ranging from basic familiarization training to front-seat advisory missions in combat conditions, much as today's teams do. The new wing will need to expand the number of aircraft available for training its pilots. Since the OAD-A teams train partner nations in how to use their own aircraft operationally, its pilots must be fully qualified in Mi-8s and -17s, AN-24s, CASA 212s, and other aircraft not found in the regular USAF inventory. 6 SOS personnel also need access to C-130s and UH-1 Hueys for training because these are used by many friendly nations. The 6 SOS is cur-

---

[45] Likewise, USAF might opt for the generic numbered special operations squadron nomenclature, create a category of operational aviation advisory squadrons, or call the new units air commando squadrons.

[46] Four squadrons would provide one squadron for each overseas regional combatant command area of responsibility (EUCOM, CENTCOM, PACOM, and SOUTHCOM). Eight squadrons would provide one squadron for each of eight major subregions—Europe, Africa, the Middle East, Central Asia, South Asia, East Asia, Central America, and South America. The squadrons might vary in number of personnel and detachments according to the demonstrated need in the assigned region.

rently assigned only two UH-1N Huey helicopters. Other aircraft are accessed via leases or charter arrangements that have been problematic, according to squadron personnel. Whatever the arrangements, the new wing will need routine, assured access to a broad range of aircraft to maintain crew proficiency. A major expansion of operational aviation advising capabilities will require a significant increase in the number of aircraft available to the wing, whether USAF owns the aircraft or makes other arrangements.

The squadrons would also conduct foreign aviation assessment and act as coordinators for higher-level advising as well. The USAF currently relies on ad hoc means to assess foreign air arms. The 6 SOS conducts some of these assessments, but most are conducted by USAF component commands, regional combatant command staffs, and various other DoD elements. Some of these organizations possess little expertise in air power or counterinsurgency operations, or in the peculiar challenges and opportunities of operating in lesser-developed countries. Establishing a wing structure for aviation advisory assistance would provide a logical umbrella for such a capability. A wing assessments group might be created, but, given the importance of language and regional cultural skills, we believe that putting this capability in the squadrons would be most effective.

A dedicated wing for aviation advising should also be able to conduct higher-level advisory missions. At present, USAF's advisory cadre focuses on tactics, techniques, and procedures. The essential objective of the operational aviation advisory missions 6 SOS conducts is the cultivation of specific and discrete capabilities in the partner air arm. The capabilities the unit cultivates might be quite basic, such as maintaining the partner air arm's Mi-8 helicopters, or might be quite complex, such as teaching nighttime tactical air assault operations. The squadron's activities are, however, limited to the tactical level. Thus, the partner air arm's need for advice on campaign or strategic use of air power in counterinsurgency remains unmet. Air campaigns are conceptually complex, all the more so in a counterinsurgency context characterized by political primacy and the need to minimize the use of force. Adding strategic planners to the squadrons would allow USAF to influence not just a partner air arm's tactics, techniques, and procedures but also its broader concept for the employment of air power. In so doing, it would capitalize on the tactical-level advisory efforts of the squadrons and help ensure the best utilization of these tactical contributions at the operational level. Higher-level advisory missions would also help with institution-building, including resource management, personnel, training, and manpower issues.

The expanded squadrons would not have to possess all the requisite expertise for higher-level advising. Rather, the squadrons would have a small number of personnel with expertise in setting up strategic planning, budgeting, logistics, acquisition, and personnel systems. These team members would assist partner nations directly in some cases. More commonly, they would reach out to regional commands and other sources of expertise (e.g., Air Force Materiel Command) to identify experts who would join them on MTTs focused on improving these higher-level capabilities. For example, strategic planners from PACAF might join an MTT for a few weeks to work with the Philippine Air Force to improve its ability to conduct joint air-ground operations or plan a counterinsurgency air campaign. Ideally, future OAD-A teams will maintain the same spectrum of specialty codes, ranks, and backgrounds current teams enjoy. Larger or more-complex missions requiring more than one OAD-A team, or missions with a very high political-military profile, would include an OAD-B team to provide more-senior command and control on the ground.[47]

A wing-level organization would provide administrative support so that the squadrons could focus on their training and advising missions. Some of the areas in which a wing could provide administrative support include training; career development; policy guidance; transportation and lodging arrangements; handling security assistance accounts; country clearance and coordination with cognizant U.S. agencies; and host-nation permissions, including status of forces and administrative access protocols. Robust wing-level administrative sup-

---

[47] In certain circumstances, very large and long missions might merit the commitment of the squadron level of command and control. Here, we are thinking of missions on the scale of the Colombian train and equip program or the reestablishment of the Afghan or Iraqi air arms.

port should allow the squadrons to increase the ratio of days deployed to personnel assigned.

Taken together, increasing the number of operational aviation advisors and the amount of institutional support will add considerable depth to USAF's aviation advisory capability. If expanded according to the three-phase approach discussed above, the force could conduct four times as many missions per year, conduct longer and larger missions, and engage the spectrum of states afflicted by insurgencies that might pose threats to U.S. national interests.

Greater Breadth. The move to a wing structure would also offer important opportunities for expanding the breadth of USAF counterinsurgency capabilities. In addition to the aviation assessments and strategic and institutional advising that the expanded squadrons could do, wing-level organizations might be established to embed advisors in partner air arms and work with Air Force Materiel Command to develop new concepts and technologies.

A wing-level structure could provide an umbrella for an embedded advisory capability. At present, USAF provides operational aviation advisory assistance by sending temporary duty (TDY) teams to the partner nation. While this mode of operation is effective in many circumstances, recent U.S. experience in Iraq and Afghanistan and longstanding British experience around the world suggest that embedding advisors in partner military units for extended periods (a year or more) offers many advantages over the TDY mode of operation.[48] Foremost among these advantages are the opportunity to accrue a more-complete understanding of the partner military organizations, the ability to shape the culture and internal workings of the partner organization, and the opportunity to monitor the implementation of U.S. assistance programs first hand to ensure that they are utilized

---

[48] This point was made prominently and repeatedly in conversations with UK Ministry of Defence planners, Whitehall, February 2005. See also Mark Malan, Sarah Meek, Thokozani Thusi, Jeremy Ginifer, and Patrick Coker, *Sierra Leone: Building the Road to Recovery*, Pretoria, South Africa: Institute for Security Studies, Monograph 80, March 2003; Martin Rupiah, "The 'Expanding Torrent': British Military Assistance to the Southern African Region," *African Security Review*, Vol. 5, No. 4, 1996; and Ministry of Defence, *Defence Diplomacy*, London, UK, Paper No. 1, undated.

efficiently. Long-term embedded advisory missions have been successful at both the individual level, with officers seconded to partner units or military educational institutions, and the group level, as epitomized by the British Military Advisory and Training Teams deployed throughout the world. A wing-level structure for USAF advisors might include an element that would provide the institutional and administrative home for embedded advisors, either individuals or groups. These embedded advisors could provide an essential complement to the TDY-based model of advisory operations and would be an excellent source of information about partner air arms.

Finally, a wing structure might include an organization tasked with helping develop new concepts and technologies for counterinsurgency operations. Some of these concepts and technologies might apply to direct USAF operations against insurgents in such places as Afghanistan and Iraq, but the organization's focus should be on concepts and technologies that are appropriate for less-developed countries facing insurgent threats. For example, in the arena of unmanned aerial vehicles, rather than monitoring the cutting edge of advanced UAV systems, the wing's concepts and technology personnel might seek simple, low-cost, and easily maintained vehicles that could fit easily into less-advanced command-and-control processes normally employed by less-developed countries. Similar opportunities might be found in rotary- and fixed-wing lift platforms; sensor payloads; strike platforms; command, control, and communications systems; and in many other areas. A dedicated concepts and technology office might identify and exploit opportunities that would be irrelevant or invisible to mainstream USAF organizations focused on cutting-edge systems for U.S. operations.[49]

Establishing a wing-level structure for operational and institutional aviation advising would therefore expand both the depth and the breadth of USAF counterinsurgency capabilities. It would allow USAF to work with more partners, more often, over longer periods, and on a broader range of operational challenges. It would also allow

---

[49] Alternatively, this activity or, more ambitiously, a battle lab for counterinsurgency, might be established at the irregular warfare center discussed above.

USAF to increase the breadth of its counterinsurgency capabilities to include higher-level and embedded advising, strategic assessments, and development of new capabilities and technologies.

Potential Impact of Expanded Aviation Advising Capabilities. A brief vignette may best demonstrate the collective effects of these proposed changes. Posit, for illustrative purposes, a hypothetical country beset by a serious insurgent movement that also poses some threat to U.S. interests.

The air commando wing could be assigned the mission of assisting the partner nation's air arm. The wing, in turn, would assign this mission to the air commando squadron responsible for the relevant region. With long-established links to the country team, USAF major commands, regional combatant commands, and SOCOM, the squadron could quickly weave itself into the overall U.S. military assistance effort. The squadron's most likely first step would be an assessment team visit to record and report on the new partner's capabilities and shortfalls at the tactical, operational, strategic, and institutional levels. Working in close coordination with the country team and other DoD elements, the squadron would use this assessment to define a program of training and advisory assistance to address the partner air arm's most important shortfalls. OAD-A and OAD-B teams would be deployed to improve the counterinsurgency tactics, techniques, and procedures of key partner air units, working in many cases with U.S. Army SOF and other SOCOM elements to cultivate integrated joint capabilities (as with the 6 SOS today). Operational advisory teams would be deployed to assist partner-nation joint headquarters, air staffs, and ministry of defense planners in developing an effective campaign plan for rolling back the insurgents' military capabilities while supporting the overall political strategy. Institutional advisory teams, built around a core of air commando squadron personnel augmented by conventional USAF subject matter experts, would deploy to help the partner air arm develop the institutional infrastructure required to sustain effective air power, ranging from recruiting and retention, through education and training, through logistics and resource management, to acquisition and procurement of materiel. This institutional advisory assistance would take full account of the local conditions, culture, and capacity to develop truly sustainable partner institutions. Continuity and depth would be provided by advisors embedded for extended periods in key partner units, staffs, and headquarters. Finally, the wing's concepts and technology office could be working to identify and/or develop innovative new concepts or technologies tailored to the specific environment of the partner nation.

The Counterinsurgency Cadre. The Air Force is an institution defined in important ways by subcommunities aligned by mission, platform, and specialty. Like any other large organization, the Air Force possesses its own internal bureaucratic politics that pit these subcommunities against one another in the contest to set the priorities for the overall institution. One of the major metrics by which these internal political contests are judged is the relative "success" (often defined as promotability) of officers belonging to different subcommunities. The better the prospects of officers in a community, the more power and influence the community is perceived to possess, and the easier it is for the community to draw the most ambitious and qualified new officers. This is certainly not limited to USAF; political scientist Stephen Rosen has argued that the acceptance of military innovations in any military institution can be judged by whether "promotion pathways" have been forged for officers associated with the innovation.[50]

By these standards, USAF's historical commitment to counterinsurgency can only be characterized as indifferent. Counterinsurgency expertise has not been a path to promotion for USAF officers. Operational aviation advising has long been perceived as a dead-end career choice. USAF abolished aviation advisors entirely for more than two decades, until the small cadre of the 6 SOS was reestablished in the mid-1990s. Today, an assignment to 6 SOS is still considered a poor career choice for an officer, even though the squadron has enjoyed considerable success in having its alumni promoted. The essential truth of the matter is that USAF's counterinsurgency cadre remains a backwater.

---

[50] Stephen Peter Rosen, *Winning the Next War: Innovation and the Modern Military*, Ithaca, N.Y.: Cornell University Press, 1991.

This must change if USAF is to rise to the counterinsurgency challenge the nation faces. As noted earlier, the personal interest and involvement of senior USAF leaders will be required if the operational aviation advisory mission is to take its rightful place among USAF's core competencies. Beyond this senior involvement, however, institutional changes will be required to alter the long-term status of the operational aviation advisory community within the Air Force.

Establishing a wing-level organization will be an important step in this regard. The wing command billet, wing staff positions, and squadron command positions will provide field-grade opportunities to officers who have opted to become part of the advisory community. Depending on the size and responsibilities of the wing, the commander could be a full colonel or, possibly, a brigadier general. Wing-level suborganizations for foreign assessments, higher-level advising, embedded advising, and concept development will also provide opportunities for quality field-grade officers in the community.

Another productive step would be to establish an identifier to be included in the personnel records of officers and enlisted personnel who serve as advisors. This identifier ought to exert a broadly positive effect on the promotability of personnel possessing it. In particular, the identifier ought to be used to select personnel for the large number of key positions USAF members fill at regional combatant commands, USAF component commands, country teams, intelligence organizations, and on the Air Staff. These tend to be senior field-grade positions, providing more potential opportunities to the advisor community. There will be a natural synergy between the expansion of USAF's counterinsurgency cadre and the services' international affairs specialist program. Personnel with advisory experience should make excellent international affairs specialists, and (potentially) vice versa. Both programs will need to be expanded and strengthened in parallel.

Finally, we should note that USAF faces twin leader-development challenges with respect to counterinsurgency. The first challenge is that of developing leaders within an expanded counterinsurgency community and giving them assignments through which they can share their expertise in the broader USAF. The most capable of these leaders should be promoted and given opportunities to command at all levels in USAF. A second challenge is raising the level of understanding of counterinsurgency and irregular warfare more broadly across the entire USAF. Although not all USAF personnel need to be counterinsurgency specialists, all will need a level of education and understanding of irregular warfare on a par with their current appreciation of conventional warfare. This is particularly important for the most senior leaders, who are unlikely to view counterinsurgency as a core USAF mission unless they have developed an appreciation for it over the course of their careers.

Enhance USAF Combat Capabilities for Counterinsurgency

The central argument of this monograph has been that the most effective means for the U.S. military to contribute to the defeat of insurgencies is indirectly, through advisory and training missions. We do recognize, however, that U.S. combat capabilities for counterinsurgency need to be enhanced for two reasons. First, as noted in Chapter Four, in some situations, there may be no alternative to sending U.S. combat forces. In such cases, the United States will want to have forces that can prevail under the unique and difficult circumstances associated with counterinsurgency. Second, U.S. advisors and trainers will not be credible with partner-nation counterparts if the United States is perceived as lacking operational competence in counterinsurgency.

For these reasons, both AFSOC and USAF general-purpose forces need improved capabilities to operate in counterinsurgency settings. Because of its mission and history, AFSOC already has some systems optimized for counterinsurgency settings. For example, because of its endurance, outstanding sensor suite, large crew, and precise weapons (especially the 105-mm gun), the AC-130 gunship has proven to be exceptionally valuable in counterinsurgency from Vietnam to Iraq. The unique design of the gunship allows it to orbit a target area and observe and fire on targets without interruption, something of enormous benefit to small friendly ground forces. USAF general-purpose forces are becoming more relevant to counterinsurgency simply through the evolutionary improvements in sensors, connectivity, and precision weapons developed for conventional conflict. For example, the advanced targeting capabilities that USAF fighter aircraft now routinely carry offer dramatic improvements in the ability to detect and identify personnel. New weapons, such as the small-diameter bomb, offer the ability to destroy a small target precisely, with limited collateral damage. MQ-1 Predator UAVs armed with Hellfire missiles offer another precise weapon appropriate for counterinsurgency settings. These systems and weapons are proving their worth in counterinsurgency operations in both Iraq and Afghanistan. Other systems (e.g., fighters or bombers carrying Joint Direct Attack Munitions or A-10s using laser-guided bombs, rockets, and the 30-mm gun) also can contribute significantly to counterinsurgency, but their use is generally limited to more-intense combat situations, particularly in rural areas.[51]

USAF combat capabilities for counterinsurgency are substantial, but significant shortfalls remain, especially in the ability of airborne sensors on platforms flying at medium altitudes to penetrate foliage, detect hidden weapons and explosives, monitor activities inside of structures, or identify personnel. Although USAF can deliver relatively small weapons with great precision, it still lacks options to neutralize individual adversaries in close proximity to noncombatants or friendly personnel, to control crowds, or to prevent movement of people on foot through complex urban terrain. A variety of sensor, aircraft platform, and weapon technologies hold great promise for counterinsurgency, but they will require a much higher level of funding and leadership support before they can be developed into practical battlefield systems.[52]

---

[51] Combat aircraft delivering Joint Direct Attack Munitions and other precision munitions provided highly effective close support for U.S. Army and Marine ground forces during the 2003 and 2004 offensive operations in the city of Fallujah, Iraq. We recognize the tactical effectiveness of these aircraft and weapons, but whether these offensive operations advanced U.S. strategic objectives in Iraq is another matter.

[52] Over the last decade, RAND Project AIR FORCE has conducted multiple studies that identified technologies and concepts for enhancing USAF capabilities against insurgents and other irregular warriors. See Alan Vick, Richard Moore, Bruce Pirnie, and John Stillion, *Aerospace Operations Against Elusive Ground Targets*, Santa Monica, Calif.: RAND Corporation, MR-1398-AF, 2001; Vick, Stillion, et al. (2000); and Alan Vick, David T. Orletsky, John Bordeaux, and David Shlapak, *Enhancing Air Power's Contribution Against Light Infantry Targets*, Santa Monica, Calif.: RAND Corporation, MR-697-AF, 1996.

CHAPTER SEVEN
Conclusions

The United States can ignore insurgencies only at its own peril. There are several reasons for this. First, the material well-being of U.S. citizens depends on the smooth running of a highly interdependent global economy. This is not just about access to raw material or cheap labor. Rather, it is the result of a global economy so tightly integrated that information flows, capital, raw materials, components, and final products must move rapidly in a highly predictable manner. Conflict, crime, and disorder, which disrupt access to or the flow of raw materials, labor, components, or finished products, are likely to have much greater economic effects than in previous years. Second, links among insurgents and terrorists with global agendas make many insurgencies potential launching pads for attacks against the U.S. homeland or U.S. interests overseas. Such insurgencies cannot be allowed to progress to the point at which they can provide sanctuary for global terrorists. Finally, to the extent that insurgencies create humanitarian emergencies (e.g., famine, genocide, massive refugee flows), the U.S. and international community may feel compelled to respond.

That is not to say that all insurgencies are the same or that the United States will need to intervene in every insurgency. Many insurgencies will present little or no threat to U.S. interests. Others can be handled by partner nations with little or no assistance from the United States. But in a few tens of countries, U.S. assistance will be essential to defeat the insurgents. Typically, U.S. advising, training, and equipping of partner-nation police, intelligence, and militaries will be the most effective means of combating insurgency. Only rarely will U.S. combat forces be called on to conduct large-scale operations.

For these reasons, this monograph has emphasized the need for USAF to develop an expanded capacity to conduct aviation advising, assisting, training, and equipping missions. As discussed in earlier chapters, police, intelligence, economic, and political advice and assistance are most important for defeating insurgents and should be given the highest and earliest priority. Nevertheless, given the importance of the military as an institution in most societies and the years it can take to make lasting advances in capabilities, early, small-scale assistance to partner-nation military forces is an important hedge against future developments and an investment that may help avoid a much more costly later intervention.

Whether as advisors or in combat operations, USAF personnel should expect to operate in a joint, combined, and interagency environment. Although air power (whether local or U.S.) will often be in a supporting role to ground forces, its contribution to counterinsurgency operations should not be underestimated. Air power is unlikely to be decisive on its own against insurgents, but it has historically proven to be enormously valuable to forces that have been fortunate enough to have access to it. Indeed, one would be hard pressed to find a commander today in any army who would choose to engage in counterinsurgency operations without substantial air support.

Final Thoughts

DoD and USAF face a great challenge in meeting these new demands and, at the same time, continuing necessary modernization to deter and defeat conventional threats. The argument that irregular warfare has completely supplanted conventional conflict is a gross oversimplification of reality. One need look no further than China to identify a nation whose economic might, ambitions, and rapid military modernization could pose a serious threat to U.S. interests in Asia. To prevail against China or other potential adversaries with advanced military capabilities, the United States must not neglect its own conventional modernization programs.

Thus, the United States must find a way to expand its competencies to deal with evolving insurgent and terrorist threats and, at the same time, maintain the ability to dominate large-scale conventional conflicts in air, land, sea, and space. Fortunately for USAF, many of its modernization programs (especially in sensor platforms, more-discriminating weapons, and data links) have great utility against both conventional and irregular opponents. The greatest institutional challenge the increasing importance of counterinsurgency poses will not be in major new weapon, sensor, or aircraft acquisition programs but, rather, in identifying and developing personnel who have the aptitude for this type of warfare and in creating organizations that can effectively advise, train, and equip partner air forces to wage internal wars that are ultimately theirs to win or lose.

APPENDIX A

States Afflicted by Insurgency

This appendix uses data from four major publicly available databases to estimate the number of states afflicted by insurgency today. The Jane's World Insurgency and Terrorism database and the Global Terrorism/Insurgency Events databases were used to compile data on the internal security situation in each of the 191 countries of the world. The Uppsala Conflict Database and the 2005 United Nations Human Development Report were used to validate the Jane's data.

The following countries are home to one or more active insurgent groups or have witnessed more than ten insurgent or terrorist events in the past two years:

Afghanistan	Ethiopia	Myanmar
Algeria	Georgia	Nepal
Angola	India	Pakistan
Bangladesh	Indonesia	Philippines
Burundi	Iran	Russia
Central African Republic	Iraq	Sri Lanka
Colombia	Israel	Sudan
Congo	Laos	Thailand
Cote d'Ivoire	Malaysia	Yemen
Eritrea	Mexico	

The following countries are home to one or more latent or residual insurgent movements or have witnessed more than one but fewer than ten insurgent or terrorist events in the past two years:

Albania
Argentina
Azerbaijan
Bahrain
Bolivia
Bosnia and Herzegovina
Cambodia
Chad
Chile
China
Cyprus
Djibouti
Ecuador
Egypt
Former Yugoslav Republic Macedonia
Grenada
Guatemala
Guinea
Haiti
Jordan
Kazakhstan
Kenya
Kuwait
Kyrgyzstan
Liberia
Madagascar
Mali
Mauritania
Morocco
Namibia
Nicaragua
Niger
Nigeria
Papua–New Guinea
Peru
Saudi Arabia
Senegal
Serbia
Sierra Leone
Solomon Islands
Somalia
South Africa
Syria
Tajikistan
Tanzania
Tunisia
Turkey
Uganda
Uzbekistan
Venezuela
Zimbabwe

Certain countries, mostly wealthy west European states, have been removed from the list despite witnessing the requisite number of events over the past two years because, in the analysts' judgment, these states face terrorist threats rather than true insurgencies.

APPENDIX B

Estimating Manpower Requirements for Advisory Assistance

In Chapter Six, we present a metric for estimating USAF manpower requirements for advisory assistance based on the experience of the 6 SOS. This appendix provides additional details on those calculations.

We derived a deployment efficiency ratio by dividing the number of days the 6 SOS was deployed (average from the years 2002 through 2004) by a theoretical maximum. Our theoretical maximum is the average number of personnel assigned to the unit during those years multiplied by 180—the USAF goal (ceiling) for days deployed per year for any individual. Thus,

$$\frac{\underset{\text{(average 2002–2004)}}{\text{Mission days}}}{\underset{\text{(average 2002–2004)}}{\text{Personnel assigned}} \times \underset{\text{(180 days TDY)}}{\text{Goal}}} = \text{Deployment efficiency ratio.}$$

Using the 6 SOS data yields

$$\frac{4{,}001}{96 \times 180} = 0.23.$$

Thus, if m is the manpower required, d is the number of mission days, and g is the TDY goal, we can shorten the equation as follows:

$$\frac{d}{g \times m} = 0.23$$

or

$$\frac{d}{180 \times m} = 0.23.$$

To solve for the manpower required to meet a specific deployment goal, we multiplied each side of the equation by 180:

$$\frac{180 \times d}{180 \times m} = 0.23 \times 180$$

or

$$\frac{d}{m} = 41.4.$$

Then, multiplying each side by m yielded

$$m \times \frac{d}{m} = 41.4 \times m$$

or

$$d = 41.4 \times m.$$

Finally, dividing each side by 41.4 yielded

$$\frac{d}{41.4} = m.$$

Stated more fully in words,

$$\frac{\text{Mission days per year goal}}{41.4} = \text{Manpower required.}$$

Bibliography

Ackerman, Spencer, "Withdraw from Iraq This Year: Save the Date," *New Republic,* February 14, 2005, pp. 17–23.

Aeroflight, "El Salvador Air Force Unit History: Grupo de Operaciónes Especiales," *World Air Forces* Web site, May 1, 2003a. Online at http://www.aeroflight.co.uk/waf/americas/el_salvador/El_Salvador-af-GrpOpEsp.htm (as of May 16, 2006).

———, "El Salvador Air Force Unit History: Escuadrón de Helicópteros," *World Air Forces* Web site, May 24, 2003b. Online at http://www.aeroflight.co.uk/waf/americas/el_salvador/El_Salvador-af-EscHel.htm (as of May 16, 2006).

Agayev, Elman, Mamuka Kudava, and Ashot Voskanian, "U.S. Security and Military Cooperation with the Countries of the South Caucasus: Successes and Shortcomings," event summary, Harvard Belfer Center for Science and International Affairs, Caspian Studies Program, May 13, 2003.

Allen, George W., "Intelligence in Small Wars," *Studies in Intelligence,* Winter 1991.

Ames, Paul, "U.S. Urges NATO to Take Role of Trainer," *Washington Post,* September 22, 2005.

Anderson, Bruce, "The Enemy Within," *Spectator* (London), October 6, 2001, p. 30.

Anderson, Jon Lee, *Guerrillas: Journeys in the Insurgent World,* New York: Penguin Books, 2004.

Applebaum, Anne, "The Discreet Charm of the Terrorist Cause," *Washington Post,* August 3, 2005, p. A19.

Asprey, Robert B., *War in the Shadows: The Guerrilla in History: Vol. I,* Lincoln, Neb.: iUniverse, 2002.

Bacevich, Andrew, *The New American Militarism: How Americans are Seduced By War,* Oxford, UK: Oxford University Press, 2005.

Bacevich, Andrew J., et al. *American Policy in Small Wars: The Case of El Salvador*, Washington, D.C.: Pergamon-Brassey's, 1988.

Bailey, Timothy, "Air Commando: A Heritage Wrapped in Secrecy," *Airman*, March 1997. Online at http://www.af.mil/news/airman/0397/comando.htm (as of May 16, 2006).

Baldwin, David A., "The Power of Positive Sanctions," *World Politics*, Vol. 24, No. 1, October 1971, pp. 19–38.

Banusiewicz, John D., "Rumsfeld Visits Georgia, Affirms U.S. Wish That Russia Honor Istanbul Accords," *American Forces Press*, December 5, 2003. Online at http://www.defenselink.mil/news/Dec2003/n12052003_200312053.html.

Barber, Noel, *The War of the Running Dogs*, New York: Bantam Books, 1972.

Bateman, Vance, "Tactical Air Power in Low-Intensity Conflict," *Airpower Journal*, Spring 1991. Online at http://www.airpower.maxwell.af.mil/airchronicles/apj/6spr91.html (as of April 25, 2006).

Beckett, Ian F. W., *Modern Insurgencies and Counter-Insurgencies: Guerrillas and Their Opponents since 1750*, London, UK: Routledge, 2003.

———, *Insurgency in Iraq: An Historical Perspective*, Carlisle, Pa.: U.S. Army War College Strategic Studies Institute, January 2005.

Beckett, Ian F. W., and John Pimlott, eds., *Armed Forces and Modern Counter-Insurgency,* New York: St. Martin's, 1985.

Bedi, Rahul, "Maoist Activity Increases in India," *Jane's Intelligence Review* (online edition), April 2005 (accessed April 18, 2005).

Bell, J. Bowyer, *The Dynamics of the Armed Struggle*, London, UK: Frank Cass, 1988.

Bensahel, Nora, *The Counterterror Coalitions: Cooperation with Europe, NATO, and the European Union*, Santa Monica, Calif.: RAND Corporation, MR-1746-AF, 2003.

Betts, Richard, "Stability in Iraq?" unclassified draft of paper prepared for the Center for the Study of Intelligence, Central Intelligence Agency, June 7, 2005.

Birtle, Andrew J., *U.S. Army Counterinsurgency and Contingency Operations Doctrine, 1860–1941*, Washington, D.C.: U.S. Army Center for Military History, 2001.

Blaufarb, Douglas, *Organizing Counterinsurgency in Thailand, 1962–1970*, Santa Monica, Calif.: RAND Corporation, R-1048-ARPA, 1972.

———, *The Counterinsurgency Era: U.S. Doctrine and Performance, 1950 to the Present*, New York: The Free Press, 1977.

Bolkcom, Christopher, and Kenneth Katzman, *Military Aviation: Issues and Options for Combating Terrorism and Insurgency*, Washington, D.C.: Congressional Research Service, January 24, 2005, p. 9.

Boot, Max, *The Savage Wars of Peace: Small Wars and the Rise of American Power*, New York: Basic Books, 2002.

Bracamonte, Jose Angel Moroni, and David E. Spencer, *Strategy and Tactics of the Salvadoran FMLN Guerrillas: Last Battle of the Cold War, Blueprint for Future Conflicts*, Westport, Conn.: Praeger Publishers, 1995.

[British] Army Field Manual, Vol. 1, *Combined Arms Operations*, Part 10, *Counter Insurgency Operations (Strategic and Operational Guidelines)*, July 2001.

Brozenick, Norman J., *Another Way to Fight: Combat Aviation Advisory Operations*, Maxwell Air Force Base, Ala.: Air University Press, 2002.

Burke, Jason, "Special Report: Al Qaeda After Spain," *Prospect* (London), May 27, 2004.

Burns, Robert, "Army Chief Says Islamic Extremist Threat is Like a 'Cancer' that Will Linger," Associated Press, June 15, 2004.

Byman, Daniel, *Deadly Connections: States that Sponsor Terrorism*, Cambridge, UK: Cambridge University Press, 2005a.

———, *Going to War with the Allies You Have: Allies, Counterinsurgency, and the War on Terrorism*, Carlisle Barracks, Pa.: Strategic Studies Institute, U.S. Army War College, November 2005b.

Byman, Daniel L., Peter Chalk, Bruce Hoffman, William Rosenau, and David Brannan, *Trends in Outside Support for Insurgent Movements*, Santa Monica, Calif.: RAND Corporation, MR-1405-OTI, 2001.

Byman, Daniel, and Matthew Waxman, *The Dynamics of Coercion: American Foreign Policy and the Limits of Military Might*, Cambridge, UK: Cambridge University Press, 1992.

Cable, Larry E., *Conflict of Myths: The Development of American Counterinsurgency Doctrine and the Vietnam War*, New York: New York University Press, 1986.

Callwell, C.E., *Small Wars: Their Principles and Practice*, 3rd ed., Lincoln, Neb.: Bison Books by the University of Nebraska, 1996 [1896].

Cann, John P., *Counterinsurgency in Africa: The Portuguese Way of War, 1961–1974,* Westport, Conn.: Greenwood Press, 1997.

Cassidy, Robert M., "Why Great Powers Fight Small Wars Badly," *Military Review*, Vol. 82, No. 5, 2002, pp. 41–53.

———, *Peacekeeping in the Abyss: British and American Peacekeeping Doctrine and Practice After the Cold War*, Westport, Conn.: Praeger Publishers, 2004.

———, "Back to the Street Without Joy: Counterinsurgency Lessons from Vietnam and Other Small Wars," *Parameters*, Vol. 34, No. 2, 2004, pp. 73–83.

Center for the Study of Civil War, The PRIO/UppsalaArmed Conflict Dataset v.3.0, 2005. Online http://www.prio.no/cwp/armedconflict (as of June 2005).

Center for Media & Democracy, *Global Struggle Against Violent Extremism*, Web site, last updated February 13, 2006. Online at http://www.sourcewatch.org/index.php?title=Global_struggle_against_violent_extremism (as of May 2, 2006).

Central Intelligence Agency, *Guide to the Analysis of Insurgency,* Washington, D.C., no date.

Chaliand, Gerard, *Terrorism: From Popular Struggle to Media Spectacle*, London, UK: Saqi Books, 1987.

Childress, Michael, *The Effectiveness of U.S. Training Efforts in Internal Defense and Development: The Cases of El Salvador and Honduras*, Santa Monica, Calif.: RAND Corporation, MR-250-USDP, 1995.

Clausewitz, Carl von, *On War*, trans. Michael Howard and Peter Paret, eds., N.J.: Princeton University Press, 1976 [1973].

Cline, Lawrence E., *Pseudo Operations and Counterinsurgency: Lessons from Other Countries*, Carlisle, Pa.: U.S. Army War College Strategic Studies Institute, June 2005.

Cohen, Eliot A., "A Hawk Questions Himself as His Son Goes to War," *Washington Post*, July 10, 2005, p. B1.

Colburn, Forest D., *The Vogue of Revolution in Poor Countries*, Princeton, N.J.: Princeton University Press, 1994.

Cole, Juan, "Zawahiri Letter to Zarqawi: A Shiite Forgery?" *Informed Comment: Thoughts on the Middle East, History, and Religion*, Web site, October 14, 2005. Online at http://www.juancole.com/2005/10/zawahiri-letter-to-zarqawi-shiite.html (as of May 2, 2006).

Conetta, Carl, *Vicious Circle: The Dynamics of Occupation and Resistance in Iraq,* Research Monograph 10, Washington, D.C.: Project on Defense Alternatives, May 18, 2005. Online at http://www.comw.org/pda/0505rm10.html (as of April 25, 2006).

Congressional Budget Office, *Estimated Costs of Continuing Operations in Iraq and Other Operations of the Global War on Terrorism*, Washington, D.C., June 25, 2004.

Conway, John L., III, "The View from the Tower of Babel: Air Force Language Posture for Global Engagement," *Air & Space Power Journal,* Summer 2005. Online at http://www.airpower.maxwell.af.mil/airchronicles/apj/apj05/sum05/conway.html (as of April 25, 2006).

Cook, Nicolas, *Diamonds and Conflict: Background, Policy, and Legislation*, Washington, D.C.: Congressional Research Service, updated July 16, 2003.

Corum, James S., and Wray R. Johnson, *Airpower in Small Wars: Fighting Insurgents and Terrorists,* Lawrence, Kan.: University Press of Kansas, 2003.

Council on Foreign Relations, "Terrorism Havens: Pakistan," Web page, updated December 2005. Online at http://cfrterrorism.org/coalition/pakistan.html (as of May 2, 2006).

Coutsoukis, Photius, "Angola as a Refuge," Web page, Rev. ed, November 10, 2004. Online at http://www.photius.com/countries/angola/national_security/angola_national_security_angola_as_a_refuge.html (as of April 25, 2006).

Craddock, Bantz, Posture Statement of Commander, U.S. Southern Command before the 109th Congress, House Armed Services Committee, March 9, 2005.

Crocker, Chester A., "Bridges, Bombs, or Bluster?" *Foreign Affairs*, September–October 2003, p. 32.

Danner, Mark, *The Massacre at El Mozote*, New York: Vintage Books, 1994.

Davis, Mike, "Planet of Slums: Urban Involution and the Informal Proletariat," *New Left Review*, No. 26, March–April 2004.

Donovan, Nick, Malcolm Smart, Magui Moreno-Torres, Jan Ole Kiso, and George Zacharaiah, "Countries at Risk of Instability: Risk Factors and Dynamics of Instability," background paper, London, UK: Prime Minister's Strategy Unit, February 2005. Online at http://www.strategy.gov.uk/work_areas/countries_at_risk/index.asp (as of May 4, 2006).

Doucette, John W., *U.S. Air Force Lessons in Counterinsurgency: Exposing Voids in Doctrinal Guidance*, Master's Thesis, Maxwell Air Force Base, Ala.: School for Advanced Air and Space Studies, 1999.

Downie, Richard D., *Learning from Conflict: The U.S. Military in Vietnam, El Salvador, and the Drug War*, Westport, Conn.: Praeger Publishers, 1998.

Downs, William Brian, "Unconventional Airpower," *Air & Space Power Journal,* Spring 2005. Online at http://www.airpower.maxwell.af.mil/airchronicles/apj/apj05/spr05/vorspr05.html (as of April 25, 2006).

Drew, Dennis, *Insurgency and Counterinsurgency: American Military Dilemmas and Doctrinal Proposals*, Maxwell Air Force Base, Ala.: Air University Press, March 1988.

———, "U.S. Airpower Theory and the Insurgent Challenge: A Short Journey to Confusion," *Journal of Military History,* October 1998, pp. 809–832.

Duffy, Michael, Mark Thompson, and Michael Weisskopf, "Secret Armies of the Night," *Time*, June 23, 2003.

Eliason, William T., *USAF Support to Low Intensity Conflict: Three Case Studies from the 1980s,* master's thesis, Maxwell Air Force Base, Ala.: School for Advanced Air and Space Studies, 1994.

Elkhamri, Mounir, Lester W. Grau, Laurie King-Irani, Amanda S. Mitchell, and Lenny Tasa-Bennet, "Urban Population Control in Counterinsurgency," unpublished paper, Ft. Leavenworth, Kan.: U.S. Army Foreign Military Studies Office, October 2004.

Fair, C. Christine, *The Counterterror Coalitions: Cooperation with Pakistan and India,* Santa Monica, Calif.: RAND Corporation, MG-141-AF, 2004.

Fearon, James D., and David D. Laitin, "Ethnicity, Insurgency, and Civil War," *American Political Science Review,* Vol. 97, No. 1, February 2003, pp. 75–90.

Feickert, Andrew, "U.S. Special Operations Forces (SOF): Background and Issues for Congress," Congressional Research Service, June 9, 2005.

Filkins, Dexter, and David S. Cloud, "Defying U.S. Efforts, Guerrillas in Iraq Refocus and Strengthen," *New York Times,* July 24, 2005, p. 1.

Finlan, Alistair, "Trapped in the Dead Ground: U.S. Counter-Insurgency Strategy in Iraq," *Small Wars and Insurgencies,* Vol. 16, No. 1, March 2005, p. 14.

Fulghum, David A., "Combat Aviators Claim 'Non-Kinetic Warfare' is Here," *Aviation Week and Space Technology,* May 23, 2005a, p. 50.

———, "USAF F-15E's New Non-Bombing Mission Draws Praise," *Aviation Week and Space Technology,* May 23, 2005b, p. 53.

Gann, Lewis H., *Guerrillas in History,* Stanford, Calif.: Hoover Institution Press, 1971.

Gearty, Conor, *Terror,* London, UK: Faber and Faber, 1991.

Graham, Bradley, "Public Honors for Secret Combat," *Washington Post,* May 6, 1996, p. A1.

Grant, Rebecca, "The Fallujah Model," *Air Force Magazine,* February 2005. Online at http://www.afa.org/magazine/feb2005/0205fallujah.asp (as of April 25, 2006).

Grau, Lester, "Guerrillas, Terrorists, and Intelligence Analysts," *Military Review,* July–August 2004.

Haas, Michael E., *Air Commando! 1950–1975: Twenty-Five Years at the Tip of the Spear,* Hurlburt Field, Fla.: Headquarters Air Force Special Operations Command, 1994.

———, *Apollo's Warriors: United States Air Force Special Operations During the Cold War,* Maxwell Air Force Base, Ala.: Air University Press, 1997.

Halliday, Fred, *Revolution and World Politics: The Rise and Fall of the Sixth Great Power,* Durham, N.C.: Duke University Press, 1999.

Hammes, Thomas X., *The Sling and the Stone: On War in the 21st Century*, St. Paul, Minn.: Zenith Press, 2004.

Hart, B. H. Liddell, *Strategy*, 2nd rev. ed., New York: Meridian, 1991.

Hashim, Ahmed S., "The Sunni Insurgency in Iraq," *MEI Perspective*, Middle East Institute, August 15, 2003. Online at http://www.mideasti.org/articles/doc89.html (as of April 25, 2006).

Heidelberg Institute on International Conflict Research, *Conflict Barometer 2004: 13th Annual Conflict Analysis*, Heidelberg: University of Heidelberg, 2004.

Herd, Graeme, and Jennifer Moroney, eds., *Security Dynamics in the Former Soviet Bloc*, London, UK: Curzon, 2003.

Hitchens, Christopher, "History and Mystery: Why Does the *New York Times* insist on calling Jihadists 'Insurgents'?" *Slate,* May 16, 2005. Online at http://www.slate.com/id/2118820 (as of April 25, 2006).

Hoffman, Bruce, *British Air Power in Peripheral Conflict, 1919–1976,* Santa Monica, Calif.: RAND Corporation, R-3749-AF, 1989.

———, *Inside Terrorism*, New York: Columbia University Press, 1998.

———, *Insurgency and Counterinsurgency in Iraq*, Santa Monica, Calif.: RAND Corporation, OP-127-IPC/CMEPP, 2004a.

———, "Plan of Attack," *Atlantic Monthly*, July/August 2004b, pp. 42–43.

Hoffman, Bruce, and Jennifer Taw, *Defense Policy and Low-Intensity Conflict: The Development of Britain's "Small Wars" Doctrine During the 1950s*, Santa Monica, Calif.: RAND Corporation, R-4015-A, 1991.

———, *A Strategic Framework for Countering Terrorism and Insurgency*, Santa Monica, Calif.: RAND Corporation, N-3506-DOS, 1992.

Hoffman, Bruce, Jennifer Taw, and David W. Arnold, *Lessons for Contemporary Counterinsurgencies: The Rhodesian Experience*, Santa Monica, Calif.: RAND Corporation, R-3998-A, 1991.

Hoffman, Frank G., "Small Wars Revisited: The United States and Nontraditional Wars, Journal of Strategic Studies, Vol. 28, No. 6, December 2005, pp. 913–940.

Horne, Alistair, *A Savage War of Peace: Algeria 1954–1962*, London: Pan Books Edition, 1997.

Hosmer, Stephen T., *The Army's Role in Counterinsurgency and Insurgency*, Santa Monica, Calif.: RAND Corporation, R-3947-A, 1990.

———, *Constraints on U.S. Strategy in Third World Conflict*, Santa Monica, Calif.: RAND Corporation, R-3208-AF, 1985.

Hosmer, Stephen T., and S. O. Crane, *Counterinsurgency: A Symposium, April 16–20, 1962*, Santa Monica, Calif.: RAND Corporation, R-412-ARPA, 1962.

Hunt, Richard A., *Pacification: The American Struggle for Vietnam's Hearts and Minds*, Boulder, Colo.: Westview Press, Inc., 1995.

The International Military Advisory and Training Team (Sierra Leone, Web site, February 8, 2005. Online at http://www.army.mod.uk/deployments/sierra%5Fleone/ (as of May 16, 2006).

Jane's Information Group, "World Insurgency and Terrorism," Web site, various dates. Online at http://jwit.janes.com/docs/jwit/browse_country.html (as of September 2005).

Jane's Terrorism and Insurgency Centre, Jane's Information Group, "Global Terrorism/Insurgency Events," Web site, various dates. Online at http://jtic.janes.com/docs/jtic/events_main.jsp (as of September 2005).

Jane's Terrorism and Insurgency Centre, Web site, http://jtic.janes.com (as of April 25, 2006).

Jehl, Douglas, "Iraq May Be Prime Place for Training of Militants, CIA Report Concludes," *New York Times*, June 22, 2005, p. 10.

Jenkins, Brian Michael, "Colombia: Crossing a Dangerous Threshold," *National Interest*, Winter 2000/2001.

Joes, Anthony James, *America and Guerrilla Warfare*, Lexington, Ky.: The University Press of Kentucky, 2000.

———, *Resisting Rebellion: The History and Politics of Counterinsurgency*, Lexington, Ky.: The University Press of Kentucky, 2004.

Jogerst, John D., "What Kind of War? Strategic Perspectives on the War on Terrorism," *Air & Space Power Journal*, Spring 2005. Online at http://www.airpower.maxwell.af.mil/airchronicles/apj/apj05/spr05/jogerst.html (as of April 25, 2006).

Johnson, David E., *Learning Large Lessons: The Evolving Roles of Ground Power and Air Power in the Post–Cold War Era*, Santa Monica, Calif.: RAND Corporation, MG-405-AF, 2006.

Johnson, David E., Karl P. Mueller, and William H. Taft V, *Conventional Coercion Across the Spectrum of Operations: The Utility of Military Force in the Emerging Security Environment*, Santa Monica, Calif.: RAND Corporation, MR-1494-A, 2002.

Johnson, Scott, and Melinda Liu, "The Enemy Spies," *Newsweek*, June 27, 2005.

Johnson, Wray, "Whither Aviation Foreign Internal Defense?" *Aerospace Power Journal,* Spring 1997. Online at http://www.airpower.maxwell.af.mil/airchronicles/apj/apj97/spr97/johnson.html (as of April 25, 2006).

———, "Ends Versus Means: The 6th Special Operations Squadron and Icarus Syndrome," *Air & Space Power Chronicles,* January 12, 2000. Online at http://www.airpower.maxwell.af.mil/airchronicles/cc/WJohnson.html (as of April 25, 2006).

Joint Chiefs of Staff, *National Military Strategy of the United States of America*, Washington, D.C.: Department of Defense, 2004.

Joint Staff, *Department of Defense Dictionary of Military and Associated Terms*, April 12, 2001 (as amended through May 9, 2005). Online at http://www.dtic.mil/doctrine/jel/doddict/data/i/index.html (as of June 5, 2005).

Jones, Seth G., Jeremy Wilson, Andrew Rathmell, and K. Jack Riley, *Establishing Law and Order After Conflict,* Santa Monica, Calif.: RAND Corporation, MG-374-RC, 2005.

Kaldor, Mary, "Iraq: The Wrong War," *Open Democracy*, September 6, 2005. Online at http://www.opendemocracy.net/conflict-ira/wrong_war_2591.jsp (as of April 25, 2006).

Kalyvas, Stathis N., "'New' and 'Old' Civil Wars: A Valid Distinction?" *World Politics*, Vol. 54, No. 1, October 2001, p. 107.

Kaplan, Robert D., "Five Days in Fallujah," *Atlantic Monthly,* July/August 2004, pp. 116–125.

———, "Imperial Grunts," *Atlantic Monthly*, October 2005. Online at http://www.theatlantic.com/doc/200510/kaplan-us-special-forces (as of April 25, 2006).

Khalilzad, Zalmay, "Priorities for U.S. Policy in Iraq," statement submitted to the Senate Foreign Relations Committee, Washington, D.C., June 7, 2005. Online at http://www.state.gov/p/nea/rls/rm/47363.htm (as of April 25, 2006).

Kilcullen, David, "Countering Global Insurgency," *Small Wars Journal*, November 2004. Online at http://www.smallwarsjournal.com/documents/kilcullen.pdf (as of October 21, 2005).

Kitson, Frank, *Low Intensity Operations: Subversion, Insurgency, and Peacekeeping*, London, UK: Faber and Faber Ltd., 1971.

Knickmeyer, Ellen, "Iraq Proposes Broader Amnesty," *Washington Post*, April 11, 2005, p. 1.

Koch, J. A., *The Chieu Hoi Program in South Vietnam, 1963–1971*, Santa Monica, Calif.: RAND Corporation, R-1172-ARPA, 1973.

Komer, R. W., *The Malayan Emergency in Retrospect: Organization of a Successful Counterinsurgency Effort,* Santa Monica, Calif.: RAND Corporation, R-0957-ARPA, 1972.

Koster, Michael C., *Foreign Internal Defense: Does Air Force Special Operations Have What it Takes?* Research Report AU-ARI-93-2, Maxwell Air Force Base, Ala.: Air University Press, 1993.

Kramer, Mark, "The Perils of Counterinsurgency: Russia's War in Chechnya," *International Security,* Winter 2004/2005, pp. 5–63.

Krepinevich, Andrew F., *The Army and Vietnam*, Baltimore, Md.: The Johns Hopkins University Press, 1986.

———, *Iraq and Vietnam: Déjà Vu All Over Again?* Washington, D.C.: Center for Strategic and Budgetary Assessments, July 8, 2004. Online at http://www.csbaonline.org/4Publications/Archive/B.20040702.IraqViet/B.20040702.IraqViet.pdf (as of May 4, 2006).

———, "Are We Winning in Iraq?" testimony before U.S. House of Representatives, Committee on Armed Services, March 17, 2005.

Laqueur, Walter, *Guerrilla: A Historical and Critical Study*, Boston: Little, Brown, and Company, 1976.

Le Billon, Philippe, *Fuelling War: Natural Resources and Armed Conflict*, Oxford: Routledge, Adelphi Paper No. 373, March 2005.

Leites, Nathan, and Charles Wolf, Jr., *Rebellion and Authority: An Analytic Essay on Insurgent Conflicts*, Chicago: Markham, 1970.

LeoGrande, William M., "A Splendid Little War: Drawing the Line in El Salvador, *International Security*, Summer 1981, pp. 27–25.

Lichbach, Mark Irving, *The Rebel's Dilemma*, Ann Arbor, Mich.: University of Michigan Press, 1998.

Linn, Brian McAllister, *The U.S. Army and Counterinsurgency in the Philippine War, 1899–1902*, Chapel Hill, N.C.: University of North Carolina Press, 1989.

Lorell, Mark A., *Airpower in Peripheral Conflict: The French Experience in Africa*, Santa Monica, Calif.: RAND Corporation, R-3660-AF, 1991.

Lumpkin, John J., "Insurgents Said to Be Infiltrating Security Forces," *Miami Herald*, October 22, 2004.

Macdonald, Douglas J., *Adventures in Chaos: American Intervention for Reform in the Third World*, Cambridge, Mass.: Harvard University Press, 1992.

Mackinlay, John, *Globalisation and Insurgency*, London, UK: International Institute for Strategic Studies, Adelphi Paper 352, 2002.

Malan, Mark, Sarah Meek, Thokozani Thusi, Jeremy Ginifer, and Patrick Coker, *Sierra Leone: Building the Road to Recovery*, Pretoria, South Africa: Institute for Security Studies, Monograph 80, March 2003

Manwaring, Max, ed., *Gray Area Phenomena: Confronting the New World Disorder*, Boulder, Colo.: Westview Press, 2003.

———, *Shadows of Things Past and Images of the Future: Lessons for the Insurgencies in Our Midst*, Carlisle Barracks, Pa.: U.S. Army War College, Strategic Studies Institute, November 2004.

———, *Street Gangs: The New Urban Insurgency*, Carlisle, Pa.: Strategic Studies Institute, U.S. Army War College, 2005.

Manwaring, Max and Court Prisk, *El Salvador at War: An Oral History*, Washington, D.C.: National Defense University Press, 1988.

Mao Tse-tung, *On Guerrilla War*, trans. Samuel B. Griffith, Chicago: University of Illinois Press, 2000 [1937].

Marks, Thomas A., *Insurgency in Nepal*, Carlisle, Pa.: U.S. Army Strategic Studies Institute, December 2003.

———, "Ideology of Insurgency: New Ethnic Focus or Old Cold War Distortions?" *Small Wars and Insurgencies*, Vol. 15, No. 1, Spring 2004, p. 111.

Marquis, Susan, *Unconventional Warfare: Rebuilding U.S. Special Operations Forces*, Washington, D.C.: Brookings Institution Press, 1997.

McCarthy, Thomas, *National Security for the 21st Century: The Air Force and Foreign Internal Defense*, master's thesis, Maxwell Air Force Base, Ala.: School for Advanced Air and Space Studies, 2004.

McCormick, Gordon, *From the Sierras to the Cities: The Urban Campaign of the Shining Path,* Santa Monica, Calif.: RAND Corporation, R-4150-USDP, 1992.

———, "A Systems Model of Insurgency," unpublished paper, Monterey, Calif.: Naval Postgraduate School, June 22, 2005.

McCoy, William H., Jr., *Senegal and Liberia: Case Studies in U.S. IMET Training and its Role in Internal Defense and Development,* Santa Monica, Calif.: RAND Corporation, N-3637-USDP, 1994.

McCuen, John J., *The Art of Counter-Revolutionary Warfare*, London, UK: Faber and Faber, 1966.

McFarlane, S. Neil, "Successes and Failures in Soviet Policy Toward Marxist Revolutions in the Third World, 1917–1985," in Mark N. Katz, ed., *The USSR and Marxist Revolutions in the Third World*, Cambridge, UK: Woodrow Wilson International Center for Scholars and Cambridge University Press, 1990.

McFate, Montgomery, "The Military Utility of Understanding Adversary Culture," *Joint Forces Quarterly,* No. 38, 3rd Qtr. 2005, pp. 42–48.

Merom, Gil, *How Democracies Lose Small Wars*, Cambridge, UK: Cambridge University Press, 2003.

Metz, Steven, *The Future of Insurgency*, Carlisle Barracks, Pa.: U.S. Army Strategic Studies Institute, December 10, 1993.

Metz, Steven, and Raymond Millen, *Insurgency and Counterinsurgency in the 21st Century: Reconceptualizing Threat and Response,* Carlisle Barracks, Pa.: U.S. Army Strategic Studies Institute, November 2004.

Middle East Media Research Institute, "The Full Version of Osama bin Laden's Speech," Special Dispatch Series, No. 811, November 5, 2004. Online at http://memri.org/bin/articles.cgi?Page=archives&Area=sd&ID=SP81104 (as of April 25, 2006)

"Military Planners Scrambling to Prepare for New Counterinsurgency Challenges," *All Things Considered*, National Public Radio, November 5, 2004.

Ministry of Defence, *Defence Diplomacy*, London, UK, Paper No. 1, undated. Online at http://www.mod.uk/NR/rdonlyres/BB03F0E7-1F85-4E7B-B7EB-4F0418152932/0/polpaper1_def_dip.pdf (as of May 4, 2006).

Montgomery, Tommie Sue, *Revolution in El Salvador: From Civil Strife to Civil Peace*, Boulder, Colo.: Westview Press, 1995.

Moore, Scott W., *Gold, not Purple: Lessons from USAID-USMILGP Cooperation in El Salvador, 1980–1992*, master's thesis, Monterey, Calif.: Naval Postgraduate School, 1997.

Morris, George C., "The Other Side of the COIN: Low-Technology Aircraft and Little Wars," *Airpower Journal*, Spring 1991. Online at http://www.airpower.maxwell.af.mil/airchronicles/apj/5spr91.html (as of April 25, 2006).

Moulton, John R., *Role of Air Force Special Operations in Foreign Internal Defense,* Maxwell Air Force Base, Ala.: Air University Press, 1991.

Mueller, Karl P., *Strategy, Asymmetric Deterrence, and Accommodation*, Ph.D. diss., Princeton University, 1991.

———, "Strategies of Coercion: Denial, Punishment, and the Future of Air Power," *Security Studies*, Vol. 7, No. 3, Spring 1998, pp. 182–228.

———, "The Essence of Coercive Air Power: A Primer for Military Strategists," *Royal Air Force Air Power Review*, Vol. 4, No. 3, Autumn 2001, pp. 45–56.

Nagl, John A., *Counterinsurgency Lessons from Malaya and Vietnam: Learning to Eat Soup with a Knife*, Westport, Conn.: Praeger, 2002.

National Intelligence Council, *Mapping the Global Future: Report of the National Intelligence Council's 2020 Project*, December 2004. Online at http://www.cia.gov/nic/NIC_globaltrend2020.html (as of April 25, 2006).

Nesbit, Roy, and Dudley Cowderoy, *Britain's Rebel Air Force: The War from the Air in Rhodesia, 1965–1980*, London, UK: Grub Street, 1998.

Norton, Richard J., "Feral Cities," *Naval War College Review*, Vol. LVI, No. 4, Autumn 2003.

O'Neill, Bard, *Insurgency and Terrorism: From Revolution to Apocalypse*, Dulles, Va.: Potomac Books, 2005.

Ochmanek, David A., *Military Operations Against Terrorist Groups Abroad: Implications for the United States Air Force*, Santa Monica, Calif.: RAND Corporation, MR-1738-AF, 2003.

Odom, William, *On Internal War: American and Soviet Approaches to Third World Clients and Insurgents*, Durham, N.C.: Duke University Press, 1992.

Office of the Director of National Intelligence, News Release, February 2005. Online at http://www.dni.gov/release_letter_101105.html (as of November 2005).

Olson, William J., "The New World Disorder," in Max G. Manwaring, ed., *Gray Area Phenomena: Confronting the New World Disorder*, Boulder, Colo.: Westview Press Inc., 1993.

Pape, Robert A., *Bombing to Win: Air Power and Coercion in War*, Ithaca, N.Y.: Cornell University Press, 1996.

Peterson, A. H., G. C. Reinhardt, and E. E. Conger, eds., *Symposium on the Role of Airpower in Counterinsurgency and Unconventional Warfare: The Algerian War*, Santa Monica, Calif.: RAND Corporation, RM-3653-PR, 1963a.

———, *Symposium on the Role of Airpower in Counterinsurgency and Unconventional Warfare: Allied Resistance to the Japanese on Luzon, World War II*, Santa Monica, Calif.: RAND Corporation, RM-3655-PR, 1963b.

———, *Symposium on the Role of Airpower in Counterinsurgency and Unconventional Warfare: Chindit Operations in Burma*, Santa Monica, Calif.: RAND Corporation, RM-3654-PR, 1963c.

———, *Symposium on the Role of Airpower in Counterinsurgency and Unconventional Warfare: The Malayan Emergency*, Santa Monica, Calif.: RAND Corporation, RM-3651-PR, 1963d.

———, *Symposium on the Role of Airpower in Counterinsurgency and Unconventional Warfare: The Philippine Huk Campaign*, Santa Monica, Calif.: RAND Corporation, RM-3652-PR, 1963e.

———, *Symposium on the Role of Airpower in Counterinsurgency and Unconventional Warfare: Unconventional Warfare in the Mediterranean Theater*, Santa Monica, Calif.: RAND Corporation, RM-3656-PR, 1963f.

Pirnie, Bruce R., Alan Vick, Adam Grissom, Karl P. Mueller, and David T. Orletsky, *Beyond Close Air Support: Forging a New Air-Ground Partnership*, Santa Monica, Calif.: RAND Corporation, MG-301-AF, 2005.

Population Information Program, The Johns Hopkins Bloomberg School of Public Health, "Meeting the Urban Challenge," *Population Reports*, Vol. XXX, No. 4, Fall 2002.

Posen, Barry R., *The Sources of Military Doctrine*, Ithaca, N.Y.: Cornell University Press, 1984.

Priest, Dana, *The Mission: Waging War and Keeping Peace with America's Military,* New York: W.W. Norton and Company, 2003.

Priest, Dana, and Josh White, "War Helps Recruit Terrorists, Hill Told," *Washington Post,* February 17, 2005, pp. A1, A6.

Project Ploughshares, *Armed Conflict Report 2004*, Waterloo, Ont., 2004. Online at http://www.ploughshares.ca/content/ACR/acr.html (as of June 5, 2005.

Pustay, John S., *Counterinsurgency Warfare*, New York: The Free Press, 1965.

Rabasa, Angel and Peter Chalk, *Colombian Labyrinth: The Synergy of Drugs and Insurgency and its Implications for Regional Stability*, Santa Monica, Calif.: RAND Corporation, MG-1339-AF, 2001.

Rabasa, Angel, Peter Chalk, R. Kim Cragin, Sara A. Daly, Heather S. Gregg, Theodore W. Karasik, Kevin A. O'Brien, and William Rosenau, *Beyond Al Qaeda, Part 2: The Outer Rings of the Terrorist Universe*, Santa Monica, Calif.: RAND Corporation, MG-430-AF, forthcoming.

Race, Jeffrey, *War Comes to Long An: Revolutionary Conflict in a Vietnamese Province*, Berkeley, Calif.: University of California Press, 1973.

Read, Robyn, "Effects-Based Airpower for Small Wars: Iraq After Major Combat," *Air & Space Power Journal,* Spring 2005. Online at http://www.airpower.maxwell.af.mil/airchronicles/apj/apj05/spr05/read.html (as of April 25, 2006).

Rice, Edward E., *Wars of the Third Kind: Conflict in Underdeveloped Countries*, Berkeley, Calif.: University of California Press, 1988.

Rich, Paul, "Al Qaeda and the Radical Islamic Challenge to Western Strategy," *Small Wars & Insurgencies*, Vol. 14, No. 1, 2003, p. 46.

Robinson, Linda, *Masters of Chaos: The Secret History of SF*, New York: Public Affairs Books, 2004.

———, "Plan of Attack," *U.S. News & World Report*, August 1, 2005, p. 26.

Rosen, Stephen Peter, *Winning the Next War: Innovation and the Modern Military*, Ithaca, N.Y.: Cornell University Press, 1991.

Rosenau, William, "The Kennedy Administration, U.S. Foreign International Security Assistance and the Challenge of 'Subterranean War,' 1961–63," *Small Wars & Insurgencies*, Vol. 14, No. 3, Autumn 2003, pp. 65–99.

———, *U.S. Internal Security Assistance to South Vietnam: Insurgency, Subversion, and Public Order*, New York: Routledge, 2005a.

———, "Recruitment Trends in Kenya and Tanzania," *Studies in Conflict and Terrorism*, Vol. 28, No. 1, January–February 2005b, p. 7.

Roy, Olivier, "Britain: Homegrown Terror," *Le Monde Diplomatique*, August 2005, p. 5.

Rumsfeld, Donald, "U.S. Refocusing Military Strategy for War on Terror, Rumsfeld Says," remarks delivered at United States Military Academy Commencement, Michie Stadium, West Point, N.Y., May 29, 2004. Online at http://japan.usembassy.gov/e/p/tp-20040601-19.html (as of September 5, 2005).

Rupiah, Martin, "The 'Expanding Torrent': British Military Assistance to the Southern African Region," *African Security Review*, Vol. 5, No. 4, 1996. Online at http://www.iss.co.za/pubs/ASR/5No4/ExpandingTorrent.html (as of October 2005).

Sayari, Sabri, and Bruce Hoffman, *Urbanization and Insurgency: The Turkish Case, 1976–1980*, Santa Monica, Calif.: RAND Corporation, N-3228-USDP, 1991.

Schelling, Thomas C., *Arms and Influence*, New Haven, Conn.: Yale University Press, 1966.

Schwarz, Benjamin, *American Counterinsurgency Doctrine and El Salvador: The Frustrations of Reform and the Illusions of Nation Building*, Santa Monica, Calif.: RAND Corporation, R-4042-USDP, 1991.

Scott, James C., "Revolution in the Revolution: Peasants and Commissars," *Theory and Society*, Vol. 7, No. 1/2, January–March 1979, pp. 97–134.

Scully, Megan, "'Social Intel': New Tool for U.S. Military," *Defense News*, April 26, 2004.

Searle, Thomas, "Making Airpower Effective Against Guerrillas," *Air & Space Power Journal,* Fall 2004. Online at http://www.airpower.maxwell.af.mil/airchronicles/apj/apj04/fal04/vorfal04.html (as of April 25, 2006).

Shafer, D. Michael, *Deadly Paradigms: The Failure of U.S. Counterinsurgency Policy*, Princeton, N.J.: Princeton University Press, 1988.

Shultz, Richard H., Douglas Farah, and Itamara V. Lochard, *Armed Groups: A Tier-One Security Priority,* Colorado Springs, Colo.: USAF Institute for National Security Studies, Occasional Paper 57, September 2004.

Snow, Donald M., *Uncivil Wars: International Security and the New International Conflicts*, Boulder, Colo.: Lynne Rienner Publishers, Inc., 1996.

Spencer, David E., *From Vietnam to El Salvador: The Saga of the FMLN Sappers and Other Guerrilla Special Forces in Latin America*, Westport, Conn.: Praeger Publishers, 1996.

Summers, Harry G., *On Strategy: A Critical Analysis of the Vietnam War*, New York: Ballantine Books, 1982.

Szayna, Thomas, Adam Grissom, Jefferson Marquis, Thomas-Durell Young, Brian Rosen, and Una Huh, *U.S. Army Security Cooperation: Toward Improved Planning and Management*, Santa Monica, Calif.: RAND Corporation, MG-165-A, 2004.

Taber, Robert, *The War of the Flea: Guerrilla Warfare in Theory and Practice*, London, UK: Paladin, 1974.

Tanham, George K., "Indicators of Incipient Insurgency," unpublished paper, 1988.

Taw, Jennifer Morrison, *Thailand and the Philippines: Case Studies in U.S. IMET Training and Its Role in Internal Defense and Development,* Santa Monica, Calif.: RAND Corporation, MR-159-USDP, 1994.

Taw, Jennifer Morrison, and Bruce Hoffman, *The Urbanization of Insurgency: The Potential Challenge to U.S. Army Operations*, Santa Monica, Calif.: RAND Corporation, MR-398-A, 1994.

Thayer, Nate, "Rubies are Red," *Far Eastern Economic Review*, February 7, 1991, p. 30.

Thompson, Robert, *Defeating Communist Insurgency: Experiences from Malaya and Vietnam*, London, UK: Chatto & Windus, 1967.

Tilly, Charles, "Does Modernization Breed Revolution?" *Comparative Politics,* April 1973, pp. 425–447.

Tomes, Robert R. "Relearning Counterinsurgency Warfare," *Parameters,* Vol. XXXIV, No. 1, Spring 2004, pp. 16–28.

Townsend, Charles, *Britain's Civil Wars: Counterinsurgency in the Twentieth Century,* London, UK: Faber and Faber, 1986.

———, *Terrorism: A Very Short Introduction*, Oxford: Oxford University Press, 2002.

Trowbridge, Gordon, "We're Rewriting the Rule Book on Counterinsurgency Warfare," *Air Force Times,* December 20, 2004.

Tyson, Ann Scott, "U.S. Pushes Anti-Terrorism in Africa," *Washington Post,* July 26, 2005, pp. A1, A14.

Ucles, Mario Lungo, *El Salvador in the Eighties: Counterinsurgency and Revolution*, Philadelphia: Temple University Press, 1996.

United Nations, List of Member States, Web page, February 10, 2005. Online at http://www.un.org/Overview/unmember.html (as of May 15, 2006).

United Nations Development Program, *Human Development Report 2005,* New York, 2005.

University of Uppsala, "Uppsala Conflict Database," Uppsala, Sweden, updated annually. Online at http://www.pcr.uu.se/database/ (as of May 4, 2006).

U.S. Agency for International Development, "USAID's Role in the War on Terrorism," Issue Brief No. 1, 2001. Online at http://www.usaid.gov/pubs/briefs/issuebrief01.html (as of April 25, 2006).

U.S. Air Force, *Foreign Internal Defense,* Maxwell Air Force Base, Ala.: Air Force Doctrine Center, Doctrine Document 2-3.1, May 10, 2004. Online at http://www.dtic.mil/doctrine/jel/service_pubs/afdd2_3_1.pdf (as of May 4, 2006).

U.S. Army, *Operations,* Washington, D.C.: Headquarters, Department of the Army, FM 3-0, June 14, 2001.

———, *Counterinsurgency Operations,* Washington, D.C.: Headquarters, Department of the Army, Field Manual—Interim 3-07.22, October 2004.

U.S. Army Center for Military History, "Vietnam Conflict—Casualty Summary," Web page, June 15, 2004. Online at http://web1.whs.osd.mil/mmid/CASUALTY/vietnam.pdf (accessed on September 9, 2005)

U.S. Army and Marine Corps, *Tactics, Techniques and Procedures for Stability Operations and Support Operations (SOSO)*, Fort Leavenworth, KS, Center for Army Lessons Learned, July 2003.

U.S. Congress, The Arms Export Control Act (as amended).

———, U.S. Foreign Assistance Act of 1961 (as amended).

U.S. Department of Defense, *The National Defense Strategy of the United States of America*, Washington, D.C., March 2005.

———, *Quadrennial Defense Review Report*, Washington, D.C.: Department of Defense, February 6, 2006.

U.S. Department of State, online report, 2000. Online at http://www.state.gov/s/ct/rls/pgtrpt/2000/2441.htm (as of December 2005).

———, "International Affairs Budget," Web page, Washington, D.C., 2006. Online at http://www.state.gov/m/rm/c6112.htm (as of May 4, 2006).

U.S. Department of State, Office of Plans, Policy, and Analysis, "International Military Education and Training Account Summaries," Web page, Washington, D.C., 2006. Online at http://www.state.gov/t/pm/ppa/sat/c14562.htm (as of May 4, 2006).

U.S. Department of State, Bureau of Intelligence and Research, "Independent States in the World," Web page, Washington, D.C., January 28, 2995. Online at http://www.state.gov/s/inr/rls/4250.htm (as of May 4, 2006).

U.S. Department of State, Office of the Coordinator for Counterterrorism, *Patterns of Global Terrorism*, Washington, D.C., April 2003.

U.S. Department of State and U.S. Department of Defense, Inspectors General, *Interagency Assessment of Iraq Police Training*, July 15, 2005. Online at http://oig.state.gov (as of July 25, 2005).

———, "Foreign Military Training and DoD Engagement Activities of Interest," Washington, D.C., 2001–2005. Online at http://www.state.gov/t/pm/rls/rpt/fmtrpt/ (as of May 4, 2006).

U.S. European Command, "Georgia Train and Equip Fact Sheet," undated [2003].

U.S. Government Accountability Office, *Security Assistance: Observations on the International Military Education and Training Program,* Washington, D.C., June 1990.

———, *El Salvador: Military Assistance Has Helped Counter but Not Overcome the Insurgency,* Washington, D.C., April 1991.

U.S. Marine Corps, *Small Wars Manual*, Manhattan, Kan.: Sunflower University Press, 2004 [1940]. Online at http://www.smallwars.quantico.usmc.mil/sw_manual.asp (as of May 4, 2006).

———, *Small Wars 21st Century,* draft manual, 2004.

U.S. Special Operations Command, *2005 Annual Report*, 2005. Online at http://www.socom.mil/Docs/2005_Annual_Report.pdf (as of May 4, 2006).

van Creveld, Martin, *The Rise and Decline of the State*, Cambridge, UK: Cambridge University Press, 1999.

Vick, Alan, David T. Orletsky, John Bordeaux, and David Shlapak, *Enhancing Air Power's Contribution Against Light Infantry Targets*, Santa Monica, Calif.: RAND Corporation, MR-697-AF, 1996.

Vick, Alan, David T. Orletsky, Abram N. Shulsky, and John Stillion, *Preparing the U. S. Air Force for Military Operations Other Than War,* Santa Monica, Calif.: RAND Corporation, MR-842-AF, 1997

Vick, Alan, Richard Moore, Bruce Pirnie, and John Stillion, *Aerospace Operations Against Elusive Ground Targets*, Santa Monica, Calif.: RAND Corporation, MR-1398-AF, 2001.

Vick, Alan, John Stillion, Dave Frelinger, Joel S. Kvitky, Benjamin Lambeth, Jefferson P. Marquis, and Matthew C. Waxman, *Aerospace Operations in Urban Environments: Exploring New Concepts*, Santa Monica, Calif.: RAND Corporation, MR-1187-AF, 2000.

Villalobos, Joaquin, "Why the FARC is Losing," *Semana* (Bogata), Foreign Broadcast Information Service, July 14, 2003.

Waghelstein, John D., "Post-Vietnam Counterinsurgency Doctrine," *Military Review,* Vol. 65 No. 5, pp. 42–49.

Walzer, Michael, "Five Questions About Terrorism," *Dissent*, Winter 2002, p. 5.

Wardlaw, Grant, *Political Terrorism: Theory, Tactics, and Counter-Measures*, Cambridge, UK: Cambridge University Press, 1982.

Warner, Margaret, "Aftermath in Spain," transcript of discussion with Richart Burt, Charles Kupchan, Daniel Benjamin, and Nicolas Checa, Public Broadcasting Service, *Newshour with Jim Lehrer*, transcript, March 15, 2004. Online at http://www.pbs.org/newshour/bb/international/jan-june04/madrid_3-15.html (as of May 2, 2006).

White House, *National Security Strategy of the United States of America*, Washington, D.C., September 2006.

Williams, Philip J., and Knut Walter, *Militarization and Demilitarization in El Salvador's Transition to Democracy*, Pittsburgh, Pa.: University of Pittsburgh Press, 1997

Wood, David, "The 'Poo Hunt': In an Unconventional War, Creative Use of Air Power," *Newhouse News Service*, August 18, 2005.

Wood, Elizabeth Jean, *Insurgent Collective Action and Civil War in El Salvador*, Cambridge, UK: Cambridge University Press, 2003.

Wooley, Michael, "America's Quiet Professionals: Specialized Airpower—Yesterday, Today and Tomorrow," *Air & Space Power Journal*, Spring 2005. Online at http://www.airpower.maxwell.af.mil/airchronicles/apj/apj05/spr05/wooley.html (as of April 25, 2006).